WILD SPACES IN
DEVELOPMENT

This fascinating book examines how microsites of spontaneous nature can reframe our understanding of the relationship between urban development and green space.

Metropolitan cities are facing stark inequalities of green space distribution, hindering goals of sustainable development. But outside of human control, spontaneous nature grows in spaces that are neglected or are unaccounted for. Drawing on existing literature and primary research in a range of towns and cities, including Quito in Ecuador, Bengaluru and Kolkata in India, and Whitby in the United Kingdom, the book delves into the morphology, meanings, and values of those small-scale assemblages of wild growth which are typically overlooked. Discussing instead how such settings can be integrated into everyday urban life, the book offers a fresh perspective on issues around green infrastructure, heritage conservation, and environmental education, enabling cities worldwide to become more nature-positive.

A unique examination of an under-researched topic, this book will appeal to students, researchers, and professionals across landscape architecture, urban planning, urban ecology, and all related fields.

Amartya Deb holds a Master of Arts in Cities and Global Development from the University of Sheffield, UK. His recent work based in India involved advising local governments on issues of sustainable urban development.

WILD SPACES IN URBAN DEVELOPMENT

Grassroots Imaginaries in a Globalising World

Amartya Deb

Routledge
Taylor & Francis Group
NEW YORK AND LONDON

Designed cover image: Amartya Deb

First published 2024
by Routledge
605 Third Avenue, New York, NY 10158

and by Routledge
4 Park Square, Milton Park, Abingdon, Oxon OX14 4RN

Routledge is an imprint of the Taylor & Francis Group, an informa business

ISBN: 978-1-032-44934-0 (hbk)
ISBN: 978-1-032-44933-3 (pbk)
ISBN: 978-1-003-37460-2 (ebk)

DOI: 10.4324/9781003374602

Typeset in Sabon
by Taylor & Francis Books

To my late father, and my mother

CONTENTS

List of illustrations *viii*
Preface *xiii*
Prologue: The first sight of a cyborg assemblage *xxviii*

1 Morphology 1
2 Meanings 21
3 Existence 41
4 Aesthetics 65
5 Collectives 89
6 Co-production 110
7 Frames 134
8 Imaginaries 156

Epilogue: Situating wild spaces in urban nature and its governance *174*
Index *182*

LIST OF ILLUSTRATIONS

Figures

0.1 Combined assessment of 50 selected sites from fieldwork using five evaluation criteria. xvii

1.1 Wild growth in a neglected space: children search for a missing cricket ball amid a patch of flowering bushes within a neighbourhood park of Bengaluru. October 20, 2019. 7

1.2 Associated with technological relic: wild plants growing around a control box for water supply along a neighbourhood street of Bengaluru. October 20, 2019. 8

1.3 Wild nature on vertical surfaces: ferns growing on exposed masonry walls of a small-scale factory in Kolkata. July 14, 2018. 11

2.1 Visual character shaped by architectural legacies: wild growth inside the locked-up bicycle stand in Bengaluru. July 26, 2015. 24

2.2 Signifying both life and death in the same scene, forces of Gaia can be felt in trees reclaiming ghosted buildings along Ramdulal Sircar Street in Kolkata. July 14, 2018. 30

2.3 Spontaneous nature as mode of greening: moss and lichens add visual relief along a narrow alley with a residential neighbourhood of Kolkata. July 14, 2018. 32
2.4 Pumpkin vines intentionally grown on a rooftop in Kolkata, show how small-scale assemblages with little or no maintenance can cater to food supply within cities. April 14, 2018. 34
3.1 Counter-utilitarian biological growth within old city fabric of Bengaluru: top – ferns growing from the ventilator shafts of a bathroom pipes; bottom – small plants growing on a *chejja* (overhang), blocking the rainwater drainage pipe. July 24, 2015. 44
3.2 Wild grass growing in empty planter boxes, and graffiti on walls of a structure in Plaza Republica of Quito. October 16, 2016. 49
3.3 Plant biodiversity in a co-created environment: a variety of wild plants growing from weathered surfaces of an abandoned building in Kolkata. The disintegration of façade has exposed wooden beams and brick masonry from colonial period as well as steel reinforcement and concrete from more recent renovations. July 6, 2018. 50
3.4 Plant growth on film soils: wild vegetation on locked footbridge of KR market building in Bengaluru, creating an appeal of a rooftop garden. July 24, 2015. 53
3.5 Unintentional forces transforming land: deep cuts in soil created by movement of vehicles have retained water after rain, to create a flash landscape within a vacant lot in Bengaluru. 56
3.6 Iron cast stairs and stone masonry create identity for an otherwise common association of plants inside Sandal Soap Factory campus of Bengaluru. January 14, 2020. 58
3.7 Growth-cycle of streetside nature: Over three-years, plant associations disappeared and re-emerged around a control box in Bengaluru. During this time, the site has survived multiple human interventions in the area. 60

3.8 Dependency of landscape fraction: the tree and blue panels belong to two sites within Bengaluru, with different ownerships. 61
4.1 Unintended nature appropriating residual space in front of a day care centre in Bengaluru. The letterings on the walls are painted in different colours to attract passer-by attention. 68
4.2 Low-tide exposes seaweed growth at the walls and foot of a seaside pier in Whitby. June 3, 2017. 74
4.3 Overgrown spontaneous vegetation within a vacant (p)lot in Kolkata, creating an impression of a forest. March 11, 2018. 75
4.4 Wild grass, construction site underway, and architectural ruins, together compose a balanced visual within Quito. October 16, 2016. 76
4.5 Boundary wall erected as visual barrier for privacy in Kolkata, now has a host of wild plant growth to become an aesthetic, ecological, as well as, cultural artefact. July 14, 2018. 79
4.6 Banyan tree next to the public water standpipe in Kolkata becomes everyday nature for people using the standpipe. July 9, 2018. 80
4.7 Coal yard within the soap factory industrial site of Bengaluru. The black soil contrasts against the otherwise predominant red soil of the region. January 14, 2020. 82
4.8 Aesthetics of post-industrial sites: spontaneous nature blends with human-built objects inside soap factory industrial yard. January 14, 2020. 84
5.1 A bouquet of plant growth on building walls of Kumartuli in Kolkata: the plaster has fallen off to expose the red brick masonry of the mercantile period. July 6, 2018. 90
5.2 Streetside nature serving as grazing patch in Bengaluru. 93
5.3 Wellbeing without Gentrification: spontaneous nature complements formal landscaping to make a neighbourhood in Bengaluru, greener. This benefits health and wellbeing of residents with no additional cost. 97

5.4 A heron searches for food in a dredged storm water drain in Bengaluru. Herbaceous plants have grown along the drain to add green to the site, imitating a pond bank. 99
5.5 Micro-habitat (a): Varying densities (red dots) of red cotton bugs along a narrow strip along the roadside in Bengaluru. 101
5.6 Micro-habitat (b): The narrow strip with red cotton bugs, in Bengaluru. 102
5.7 Wild climbers with berries growing on park fence in Bengaluru. Berries, and other fruits, if edible, may become food for birds and animals. 104
6.1 Premature stage of co-production: giant stranglers growing on top of an abandoned warehouse in Kolkata. March 31, 2018. 113
6.2 Gradients of managing spontaneous landscapes in a public park of Bengaluru: wild growth in its full bloom offers a unique aesthetic for visitors. 116
6.3 Removing ground cover: the verges of a street in Bengaluru which earlier served as grazing grounds and sponge spaces are now fully paved. The streetscape looks "clean" and "new", and saves on maintenance trading off one stratum of biodiversity. c. 2021–22. 117
6.4 Community management of wild spaces: a privately hired gardener trims an overgrown patch of wild plants that had blocked the entry of a residence in Bengaluru. 122
6.5 Biofilm in an effluent treatment tank within the soap factory of Bengaluru: this setting has offered ecologically stable conditions for growth of algae and become habitat for fish, and insects. January 14, 2020. 123
6.6 Nature integrating with a road within Whitby Abbey grounds. June 3, 2017. 125
6.7 Competition for space: mosaics formed as a variety of herbaceous plants (or ground cover) are gradually taking over a road in Whitby Abbey. June 3, 2017. 126
6.8 Exemplar precedent of small-urban places: even if lacking space, residents have grown plants inside planter boxes in Kolkata. July 15, 2018. 128

6.9 Bricks laid to form a path over wild vegetation in a front yard of a residence in Kolkata. April 14, 2018. 130
6.10 Comparing maintenance levels of two sites in Bengaluru: site A has a broken control box with litter around it; site B has a control box that is intact with wild flowers growing around it. c. May–June 2020. 131
7.1 Framed between textures and planes: vegetation growing near a public standpipe in a street corner of Kolkata. April 14, 2018. 137
7.2 Plants acting as repoussoir: stranglers grow on either side of a building façade in Kolkata. July 9, 2018. 139
7.3 Backdrop in contrast with landscape: a variety of plants grow in front of a maroon-painted wall in Kolkata. April 14, 2018. 140
7.4 Context as risk: trees growing on the roof of buildings within a residential neighbourhood in Kolkata. Some buildings here are inhabited, while others are vacant. July 9, 2018. 143
7.5 Context depreciating art value: a roadside garbage collection point in front of a unique assemblage of wild plants and buildings at Amherst Street of Kolkata. July 14, 2018. 147
7.6 Aerial view of the control box in Bengaluru, islanded by wild growth around it. June 20, 2020. 149
7.7 Nature growing inside corner of an open stormwater drain in Bengaluru. August 11, 2018. 150
8.1 Machines as part of post-industrial landscapes: a model-helicopter kept on a street side verge in Bengaluru, blends with wild plants underneath, and around it. September 27, 2018. 164

Table

6.1 Evaluating the exemplar precedent as an urban place 129

PREFACE

State of nature in cities

Urban environments face the challenge of catering to the lives of billions of people who are to inhabit cities in the future. Nature in cities, especially in the Global South, is in a steep decline, unable to compete for the limited land and available water. Metropolitan cities such as Hong Kong, Hyderabad, Mumbai, and Shanghai in Asia, or Buenos Aires and Chihuahua in Latin America, or Addis Ababa in Africa suffer from one common ailment – they have far less[1] green space per person than the WHO prescribed benchmark of 9 sqm (see also Gelan & Girma, 2022). There are, of course, other cities that have managed to increase their green cover far above the WHO benchmark, yet the disparity in distribution of urban forest cover within cities is found both in the Global North and Global South (Gerrish & Watkins, 2018; Rigolon, Browning, Lee, & Shin, 2018); thus signalling inequalities of access to nature as a planetary phenomenon of urbanisation. This lack of green cover makes cities less liveable.[2]

Plants serve as a life-support system for both human and non-human life. Yet, city governments often find it challenging to allocate resources required for maintaining landscapes, especially in poor neighbourhoods. As a result, greening projects such as tree planting do not happen; parks remain neglected; and planter boxes along the street never realise their horticultural potential envisaged by city decision makers. But while planned efforts fail, small quantities of nature may be found growing on the walls and roofs of residential dwellings, amid a marketplace, or in the corner of a park, and so on. They make for some interesting landscapes within cities. However, these novel arrangements can remain unnoticed given their small scale and low status, or are hidden from our regular experiences of the city.

While green spaces are being planned to make cities more liveable, the existence and governance of spontaneous nature in cities is often less discussed as they may be considered as transient nature and as symptoms of urban failure. Uncertainties of risks to human life also prevent architects, conservationists, and biologists from exploring alternative purposes of such nature; limiting potential for creativity and science. The lack of recognition and poor management of these sites is a missed opportunity for delivering quality urban green spaces for urban citizens; especially in neighbourhoods where lack of resources constrain the delivery of green spaces.

Scope of the study

Outside of human control, nature grows in spaces that are neglected or are unaccounted for in cities. Spontaneous nature grows not only in parks, on open ground, or people's backyards, but also on microsites such as a hole in the wall or a small patch around a street-side electric pole. In the broad visual of urban sceneries – made by skylines and streets, retail shops and rickshaws, pillars and pedestrians – one can find trees, shrubs, and grasses growing in small quantities, often creating tiny green dots on the otherwise vast grey urban landscape. Despite their common existence, very little research has been done to understand and integrate these natures within our urban spaces. This book attempts to bring out characteristics of plant-built arrangements to advance knowledge about small-scale spontaneous nature in cities.

Urban forests by definition consist of all trees, shrubs, and grasses in the city.[3] Yet, the focus of research has been on large green spaces – both formal and informal. It remains a question as to how smaller patches of nature contribute to the health of the urban forest and wellbeing of people in the city. This study of wild spaces particularly focuses on microsites of nature, so as to fill the aforementioned lacuna in the literature. Studying spontaneous nature also offers an opportunity to understand values of nature in spaces that are otherwise neglected. In addition, the book offers examples of how cities can aim for maximising benefits of small-scale assemblages of wild nature.

The prevailing frameworks of addressing urban nature include terminologies such as urban forest, green infrastructure, and park systems. These overarching conceptualisations sometimes include wild landscapes, but are arguably restricted to large settings such as wetlands and riparian vegetations. Broadly speaking, "top-down" structures of imagining nature within cities skew the importance of vegetation growing in microsites. On the one hand, I refer to acculturation of globally appealing grand visions that have typically rendered aesthetics of locally formed wild settings as undesirable. On the other hand, I also refer to units of analysis emerging from the use of powerful representations of space (e.g., architectural drawings, satellite imageries, town planning maps, and urban design plans) – drawing the attention of researchers in landscape studies to large urban spaces (e.g., woodlands, parks, backyards, or vacant lots); yet restricting the enquirer from delving into smaller fragments of nature, that nevertheless continue to exist. To overcome this limitation, in this book, I consider a bottom-up approach to understand urban nature; by focusing in particular on small-scale informal green spaces found in the city.

What is a landscape fraction?

The unit of analysis in this book, or the term *Landscape Fraction*, refers to an (aesthetic) assemblage composed by visually breaking down a larger landscape into its smaller parts. Typically, the word *landscape* is associated with vast tracts of land, as far as our eyes can see: we recognise them as meadows, gardens, forests, mountain ranges, deserts, and so on. But in densely built cities, assemblages of nature more often than not cannot be separated from the context of built environment within which they grow. In cities, where space is scarce, nature outside human care has had to adapt to harsh conditions giving rise to novel arrangements. In the cities of Whitby (in the UK), Quito (in Ecuador), as well as Bengaluru and Kolkata (in India), unique and potentially aesthetic ensembles have formed made of wild plants growing on human-built surfaces like buildings, roads, and compound walls. The word *fraction* refers to only a tiny portion of this urban landscape, around small-scale assemblages of nature. Interesting compositions of built environment and nature, often blur surrounding elements of the larger urban landscape – creating smaller frames through which one is able to see and judge the assemblages. The definition of the word *landscape*, in a broad sense, is well known as an area perceived by people. But I add the word "fractions" after landscape as aesthetics in my chosen settings are created by compositions of smaller elements rather than their whole. *Landscape fractions* are then, in a sense, landscapes within landscapes.

To identify strategies, for appropriating and programming, even the smallest residual spaces in cities so as to contribute to urban development, the study brings out the value of small-scale assemblages of nature. Here my own observation is the main instrument of analysis; and it is rather easy for prejudice to seep in. To avoid bias during data collection, I documented a wide range of sites with spontaneous nature, even when they did not appeal to me as "valuable" or "aesthetic". The search for landscape fractions in cities started with botanical growths on buildings that were rather unsettling – and yet interesting. Later, based on findings during fieldwork, the inquiry was extended to plant growths on technological relics, residual spaces, and other neglected spaces.

Then, what is broadly referred to in this book as assemblages of landscape fractions, should have one or more of the five characteristics of wild spaces below:

- **Associated with buildings:** plants had grown attached to buildings. E.g., trees growing on rooftops, or vines climbing up the pillars, or moss growing in moist niches of walls. These compositions would fit the description of natural growth reclaiming land taken away from nature by construction of buildings.
- **Associated with technological relics:** nature had grown attached to human-made objects other than buildings. E.g., vines on electric transformers or on abandoned furniture and automobiles.
- **Small-scale:** typically, sites had one or more strands of plant species. While some researchers have assigned a size to microhabitats of nature, this book has been flexible. Imagine a <1 sqm patch of moss growth or 2 sqm patch of grass growing around a control box along the roadside, or just the portion of the wall where the tree has grown rather than the whole building.
- **Residual spaces:** plants had grown in undefined and left-over spaces. E.g., below elevated water tanks, corners of a street, and so on.

- **Other neglected spaces:** wild nature had grown in spaces such as parks and backyards as a result of oversight in maintenance from managers and owners.

Based on the five characteristics above, a personal assessment with discursive scoring across a set of about 50 selected sites from my field-work – most of which are featured within this book – revealed that the majority of assemblages that had drawn my attention, can be described as small-scale nature associated with buildings (Figure 0.1). This is perhaps not surprising, when parts of urban areas are increasingly transforming to become densely built, leaving little or no space for wild plants to grow on the ground; urban soils are made impervious through paving to save costs of maintaining ornamental landscapes or to control the growth of spontaneous or "unplanned" vegetation. Thus, with little medium for growth and intense human control on ground, the building walls and roofs are where nature ends up growing. Plants also grow in semi-covered storm water drains, and other niches – hidden from our experiences of everyday nature. In the set of exhibits within the book however, there will be some which would deviate from plant-building typology of assemblages – into other spaces of neglect which are not necessarily residual spaces.

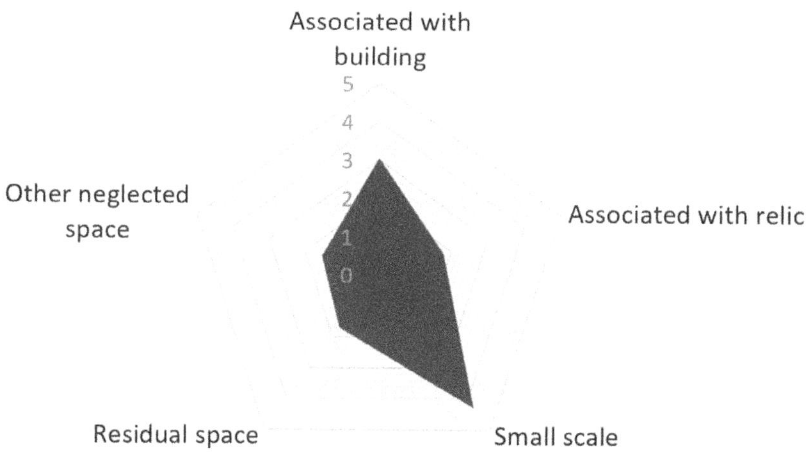

FIGURE 0.1 Combined assessment of 50 selected sites from field-work using five evaluation criteria.

Interpretations of assemblages

While small-scale spontaneous nature grows just about everywhere, wild spaces in old city areas, derelict industrial districts, and low-income neighbourhoods are particularly interesting: they may integrate with the built fabric and lives of the people in manners that add character in terms of historicity, visual-relief, food for livestock, micro-habitats, and so on. Wild urban spaces may hold secrets waiting to be discovered, as in the case of, say, ensembles of machines and plants found in industrial yards. Such assemblages may comprise of relics that would seem completely out of their usual place; making these assemblages appear strange and illogical. Such novel settings can facilitate new imaginations, and possibly encourage a greater connection of urban dwellers with nature.

Old city areas are often compact, with little space on the ground for plant growth. However, the buildings themselves offer a fertile medium for growth; and exposed masonry is a plus for anchorage of plants. Despite their heritage values, old buildings may be considered obsolete for more recent purposes and owners may be willing to spend less, with investments planned to demolish these and make way for new buildings. At other times, houses may be abandoned because of poor structural condition or the outmigration of residents. Meanwhile, in the low-priority areas – often synonymous with low-income areas – there may be neglect from public authorities to plant and maintain designed landscapes. As a result, roadside verges and parks no longer remain products of human-conceived beautification but become grounds for reclamation by nature.

Small-scale assemblages of wild growth can have both positive and negative meanings associated with them at the same time; much as it would be with spontaneous nature in general. Sometimes, despite negative contexts such as risk to life, urban decay, and dilapidation of structures, observers are able to identify landscape fractions as aesthetic settings with before and after pictures of interesting assemblages on social media. Other times, negative meanings override positive ones. Such conditions prevent a fuller evaluation of assemblages. For instance, biologists may refrain from researching for habitats on structures that are dilapidating and have become a risk to life. More importantly, when negative connotations become generalised, they stigmatise spontaneous growth; for instance, people may think of all informal growth as allergy causing or as weeds. The risks nevertheless are real in many cases.

Having said that, in some cases, cities have been able to integrate such assemblages productively, through monitoring and management.

Management regimes of spontaneous nature

Insights from studying landscape fractions reveal how people have responded to small-scale wild spaces in cities. They bring out several ways in which wild growth can be managed with minimal intervention to create tactical opportunities for communities:

- **Grazing:** Small-scale spaces of spontaneous nature are used as grazing spots for cattle. Spaces include patches of grass growing in the corners of a private plot or along roadside verges owned by the government. Such sites are highly fragmented but form a matrix as the street networks connect small patches within a neighbourhood to larger patches of spontaneous grass beds in neglected public parks and in unguarded (or even abandoned) private plots. *At times, this leads to over-grazing, and other times, it controls biological growth.*
- **Place-making:** Sometimes private property owners without many resources to spare have responded to spontaneous nature in their yards by integrating them as landscapes by minimal additions (e.g., pathways). *This maintains naturalness of the landscape, and offers better quality of regulatory ecosystem services.*
- **Controlling growth:** Spontaneous growth has been retained in areas – out of deliberate neglect, or by periodically trimming – where people find it an acceptable addition to aesthetic and ecological quality of the built environment. *Thus, wild nature becomes the primary means of greening and adds to better health and well-being of communities.*
- **Ensuring safety:** Natural growth on buildings has been managed to mitigate risks of structural collapse from the increasing load of plants or as protection from a harsh climate, through more sophisticated and calculated engineering measures. *This helps conserve heritage structures and preserves collective memories of local communities.*

Co-production of urban space

Wild spaces in cities are created through the process of plants taking over surfaces of the built environment. Although both human and natural forces are involved, the intention of human agency to integrate these novel landscapes back into urban environments is often missing – thus rendering the assemblages as partly coproduced. While some grassroots imaginaries emerge from community-led appropriation of wild spaces to demonstrate how more mature forms of co-production can nevertheless be achieved, there is a need for a more organised approach. To convert wild spaces from a waste to a resource, undertaking the following may be considered:

- **Landscape management strategies** can help enhance benefits of wild spaces. Some communities have identified wild spaces for foraging food and flowers, while others have accepted them as front yard gardens.
- **Documenting interesting sites** where small-scale nature can be preserved or encouraged in cities, can help turn marginal spaces into resources. In doing so, one should bear in mind the heritage of built elements and ecosystem services of natural features that lead to collective impact of small-scale landscaping.
- **Inspection by engineers** can help assess, and where possible, prevent risk of structural failures due to trees growing on buildings.
- **Assigning as grounds for educational and cultural activities** by integrating wild spaces with community activity networks for tourism, production of creative works (e.g., films), scientific activities, and hobbies related to exploring nature.

Organisation of chapters

The book is organised topically. Some chapters describe landscape fractions in terms of what they are and how they are assembled, while others help understand values and offer alternative imaginaries around such ensembles of nature in cities.

The prologue takes the reader back in time, to one of the first moments when I thought of plants growing on buildings as an object of interest. Using the example of a site in Bengaluru's old city area, I introduce a typical setting of a "cyborg assemblage". The term is discussed in more detail, later in the book, yet this should paint a mental picture of the objects under enquiry. Hopefully the dramatic opening would also help in the understanding of the urban landscape for readers who are perhaps more familiar with botany than urban planning.

The first chapter on Morphology sets the scene by positing spontaneous nature as part of the urban forest; and probes the inquiry for small-scale nature's role and place in cities. Using examples from fieldwork (in Kolkata and Bengaluru), I unpack the idea of landscape fractions as "small-scale assemblages" and as "landscape within landscape". The chapter points to where assemblages of enquiry may be found offering a level of certainty. It delves into descriptive analysis of their compositions; and mentions, in particular, early experiments in Germany and Japan to classify urban biotopes. The chapter ends by presenting how our top-down understanding of urban nature is falling short of delivering adequate green spaces; and how localised settings are more grounded realities of cities.

Chapter 2 on Meanings deepens the link between urbanisation and landscape fractions. It connects to concepts of urban blight, cyborg urbanisation, low-income neighbourhoods, etc. This should ground the study in the more local contexts of assemblages, and depart from the generalised imaginaries of urban nature such as avenues, parks, or playgrounds. In doing so, this chapter presents a set of possible meanings, that make landscape fractions a product of social construction; such constructions are often multi-layered. In particular, it argues that the plant growth and built elements offer new meanings which if separated, they wouldn't. Furthermore, it explains how the isolation from a city's metabolic processes, makes it difficult to value spontaneous nature growing on buildings as green infrastructure; even when they are the only mode of greening in certain urban neighbourhoods challenged with scarcity of resources such as land, water, and funds.

While Chapters 1 and 2 discuss where landscape fractions can be found and what they look like, Chapter 3 on Existence delves into how these plant-built compositions come into being, and continue to survive. The main argument in this chapter is that spontaneous growth challenges the notion of cities as human-dominated systems. Factors which are human-driven, yet beyond human control, enable conditions for plants to grow and form the assemblages – these I call sites of convenience. Convenience can be created through social (e.g., migration), economic (e.g., unemployment), political (e.g., austerity), and natural (e.g., wind) factors. Built structures give identity to plants growing around them (e.g., an industrial setting). This chapter also highlights how built-nature assemblages are unique and that such assemblages need not be temporary (as is largely perceived) but in many cases seem to stand the test of time, against complex urban processes that threaten their existence.

Given the centrality of visual character for considering an assemblage as a landscape fraction, it warrants a dedicated chapter. Chapter 4 on Aesthetics builds a vocabulary to study the language of assemblages. Broadly, three typologies emerge: architecture heavy, nature heavy, and balanced. The chapter argues that spontaneous nature can be beautiful, and looks beyond their form into aspects such as heritage, and landscape experiences which they evoke. Assemblages as discussed before are linked with time and memory, and are thus comparable with monuments. Industrial legacies – now becoming redundant in many cities – have become grounds for unique ensembles of built objects and nature. Some industrial sites have been experimented with, to be musealised as public spaces.

Chapter 5 on Collectives looks at how these fractions can be visualised as "inter-connected networks". In doing so, it highlights the roles of these "networks" in delivering ecosystem services. These benefits are more than what immediately meets the eye. Along with concepts mentioned in earlier chapters, this chapter introduces a few additional themes – *inter alia*, eco-gentrification, urban grazing, climate resilience of cities – to effectively answer the fundamental question raised in Chapter 1 on "what is the role of small-scale nature under the broad umbrella of urban forest?" This chapter should convince the reader about the values of landscape fractions with concrete evidence. At the same time, it would hopefully open up minds towards future research on how else small-scale nature is integrated with the city; culturally, ecologically, and economically.

Chapter 6 on Co-production unpacks a phenomenon, where urban dwellers have managed to purposefully integrate wild growth into the urban landscape. In other words, *co-production* is humans making use of sites which nature has reclaimed, without destroying the novel assemblages of wild growth in the process of adaptation response. Within cities, humans have sometimes retained these assemblages by creatively adding to them, or by managing them, to make them a part of everyday life; although such responses are rare occurrences. More commonly, small-scale spontaneous nature disappears in cities because assemblages are not integrated in designs of urban spaces. As alternatives to discarding assemblages, they can be turned into small urban spaces through adding design elements like pathways; or through prevention of littering; or managing biological growth.

Chapter 7 on Frames focuses on "how to" analyse landscape fractions. The idea of a frame is discussed as two-fold: first, the physical features of the site that guides the human eye – these happen through backdrops, creating vistas, enclosures, borders; second, the discursive context (e.g., is it a photograph on a magazine cover? Is the area around the fraction vandalised?). Meanings can change when the wider context of aesthetic precincts are taken into consideration (e.g., depreciate visual appeal by blighted surroundings or urban decay). Furthermore, plants can also have a "hidden life", and nature fractions can be found tucked away in drains and street corners, concealed from the sight of the passer-by. This chapter ends with how documentation is important, as it can help magnify these sites to bring their existence into people's notice.

Chapter 8 on Imaginaries is the concluding chapter, and knits together findings in the book by putting them into four broad perspectives that characterise small-scale spontaneous nature in a balanced, yet constructive way. These are: 1) ambivalence, 2) experimentation and discovery, 3) reconstruction of meanings, 4) shifting status of wild growth. These four imaginaries cross-cut learnings from all previous chapter themes. They synthesise insights to reflect, as to what extent, previous works by eminent researchers on urban nature are applicable to small-scale growth.

The epilogue of this book is a brief recap, and a call to action. It stresses what can be learnt about cities from spontaneous growth of nature: reiterating a more grounded view of urban green spaces. It positions wild spaces as an opportunity for governance of urban nature, and emphasises how important immersive experiences of wild spaces can be for urban dwellers e.g., for getting a first-hand understanding of ecological succession that cannot happen with highly manicured and grafted landscapes, which pop-up overnight. The appeal to urban futures is for wild spaces to be evaluated before being discarded – if appropriate, may also be treated, for reaping benefits.

About this book

When the aeronaut looked into the sky, she saw nothing, yet discovered a complex world of atmospheric science. Then how could it be possible that assemblages which were visible to my eye had no significant meaning? Such is the genesis of my inquiry on small-scale spontaneous nature in cities. In the summer of 2018, I received a photograph over the cellphone from my father, in his effort to persuade me to visit him in Kolkata. "These are everywhere", he said. "Why don't you come and see?" Upon my visit to Kolkata, I truly found the city replete with sites of spontaneous plants growing out of the buildings. Until now, I had been collecting pictures of wild urban spaces from cities across Ecuador, India, and the United Kingdom, thus the visit to Kolkata only added to my already existing library of resources. Soon after, I realised that I had amassed data from cities that are not commonly discussed in existing literature on wild spaces. Although the topic of spontaneous urban nature has not been conducted much in research, it has evoked some interest from urbanists, naturalists, landscape ecologists, among others. This book draws inspiration from works on wild growth in cities by several distinct voices including but not limited to – Anna Jorgensen, Jennifer Wolch, Matthew Gandy, Peter del Tredici, and Timothy Beatley. While their works span across vast lengths and breadths, I have managed to touch upon their knowledge, as much as the tip of an iceberg, and in a way that only benefits my study. While I am honoured to have drawn from works of several academics who have laid foundations on the subject, I am indebted to Matthew Gandy for encouraging my thoughts and ideas over the past four years, for bearing patience with my faults, and of course, for inspiring me to explore the depths of urban nature.

The recently completed project *Rethinking Urban Nature*, funded by the European Research Council, led by Matthew Gandy at the University of Cambridge, already has several rigorous publications which analyse wild settings of urban nature through the lens of urban political ecology and other approaches. Resonating with these themes, works of Donna Haraway, Erik Swyngedouw, and others have helped shape many ideas in this book. Of the memorable things that happened in relation to writing this book, I have been particularly fortunate to have discussed my thoughts with Bruno Latour, during a workshop on *Rediscovering Soils*, at the University of Sheffield. Rather inevitably, directly or indirectly, my inquiry has been influenced by several works of Henri Lefebvre – *The Critique of Everyday Life, Production of Space*, and *Urban Revolution*. In addition, Michel Foucault's book, *Order of Things*, served as window to the works of Bonnet, Buffon, Tournefort, Cesalpino, Crollius and others on nature; which, needless to emphasise, drove my imagination to a great extent.

This book, *Wild Spaces in Urban Development*, is broadly for those who are involved with imagining the future of cities. It is especially aimed at making architects, urban planners, and designers think about how small-scale spontaneous nature can be integrated into urban placemaking to shape sustainable cities. I remain thankful to Nicola Dempsey and Ross Cameron at the University of Sheffield, for having pointed me towards several key debates in landscape architecture. In addition to academic publications on spontaneous urban nature, I also draw from writings of practitioners that include, but are not limited to, architects, artists, photographers, and surveyors who have engaged with assemblages of human-built and spontaneous nature in their professional assignments.

It occurred to me that perhaps the reason we have so far, and largely been looking at negative constructs of spontaneous nature in cities – is because of the kind of sites that have been documented and represented as urban nature. So, expanding this dataset was one of my objectives. Between 2015 and 2021, I documented hundreds of sites with spontaneous nature in cities. Photography offered an opportunity to capture ensembles which appeared beautiful to my eye – hoping that they would hold potential in adding constructively to the current discourses of spontaneous nature. Photographic documentation also helped me apply a range of research tools (such as photo-elicitation, content analysis of photographs, and re-photography) to study urban nature and communicate findings more strongly. Carrying out planned and unplanned transect walks were helpful to maintain continuity in experience of the urban fabric in which landscape fractions were situated. Some sites, which I had regular access to, allowed me to observe and record change over time. Since I draw examples from cities I visited, some insights are from my lived experiences with communities for whom these landscapes are an integral part of everyday life. I believe anyone with this exposure would easily notice there to be a connection between plant growth, built structures, and the context of spaces in which combinations of these were assembled.

Some challenges were faced. One being that, due to limited time available during fieldwork, I was unable to acquire detailed information about some of the sites. In this case, most probable assumptions were made by looking at the pictures. This limitation should not bother the reader too much, as a building's purpose – of say as a warehouse or not – would make little difference to the overall critique of plants growing on the building. Moreover, assumptions used to fill in missing pieces of information were made from experiences of similar sites elsewhere. Another particular difficulty was to find the age of buildings or relics – abandoned buildings were particularly problematic. So, it would be prudent for the reader to treat the examples used in this book as illustrations rather than as historic documentation.

Over the past four years, my understanding of the topic has improved, thanks to the discussions with more people than I can possibly mention. The discussants included landscape architects, urban designers, urban planners, biologists, conservation architects, business management and finance professionals, economists, and geographers among others. Some discussions were structured as interviews; others were spontaneous with the freedom to meander and explore the topic. It goes without saying, that both kinds of conversation helped equally, and I thank the discussants for their time. The more structured discussions included photo-elicitation that allowed an in-depth exchange of small-scale wild spaces in cities. The chapters within the book, of course, duly credits the discussants where such insights have been drawn from.

The book mainly draws field insights from Indian cities, in addition to my brief experiences of cities in Ecuador and the UK. Their contexts are vastly different. I have lived in Bengaluru (formerly Bangalore) long enough to call it my home, and yet I owe colleagues at the World Resources Institute (WRI) Ross Center for Sustainable Cities, especially Robin King and Jaya Dhindaw, gratitude for much of my understanding of the city has evolved from working with them over the years. One of my study sites in Bengaluru, is the Sandal Soap Factory, and I extend my gratitude to both the senior executives and officers of the Indian Forest Service at the establishment, who have welcomed me into a site otherwise restricted to the public. I am especially thankful for having been generously offered an elaborate guided tour of the otherwise highly guarded industry campus.

Although the city of Kolkata (formerly Calcutta) is my birthplace, I grew up in the countryside, quite detached from the city life. Thus, I had arguably very little knowledge of the city before I started working on this book. To this end, I was fortunate to have found family and kin offering me the much-needed time to help me understand the layers of the city. Over the course of about half a year, the long walks through the streets of the old city, and candid interactions with residents, painted a vivid picture of post-colonial legacies. During my short stay in Kolkata, I also had the opportunity to attend seminars from scholars – both from universities in Kolkata and around the world – on the history and heritage of the city, and learned from them greatly.

The book draws several examples from Europe; particularly from the UK where I spent a year, pursuing my MA from the University of Sheffield in South Yorkshire – not very far from the town of Whitby in North Yorkshire. I am thankful to my dear friends Rachel Carrington in the UK, and Stefan de Koster in the Netherlands to have offered me insights of planning and everyday life practices that I had not encountered in mainstream literature.

During the brief visit to Ecuador in 2016, my primary role was that of a policy analyst on behalf of the University of Sheffield at the Habitat III conference of the United Nations. Over the few weeks there, I was able to spend time walking the streets of Quito – which was a rich learning experience in itself. There was not only ample opportunity to document microsites of nature, but to witness the management of the first city to be declared (in 1978) by UNESCO, as a World Heritage Site.

This project would never have materialised without the timely advice of Atiya Habeeb Kidawi, based in Delhi, India. I wholeheartedly extend my gratitude to anonymous reviewers, who I surmise were senior researchers specialised in the subject, from the UK and the USA. Their kind yet relevant comments have been instrumental in the development of this work. Finally, I thank the editorial team at Routledge – Russell George, Alanna Donaldson, Nick Craggs, Evie Lonsdale, and Jake Millicheap who are based in Oxfordshire, the Yorkshire Dales, and London in the UK, along with Kate Schell and Megha Patel based in New York and Tennessee in the USA – for having shared my enthusiasm related to the importance of this work.

Whereas my quest for writing this book stemmed from a desire to understand meanings of an otherwise overlooked class of urban nature, my curiosity was bolstered when I found out how little has been written across the world on people's everyday interactions with wild spaces in the city. While it is hard to suggest the next steps for the city (as I do not know how many or who would share the interest of these settings with me), I expect urban planners, urban designers, and architects to use insights for placemaking and for bottom-up planning of green spaces. I anticipate that organisations working with heritage conservation might decide to incorporate the stories of built-nature assemblages into their transect walks. Or perhaps, more cellphone apps will be designed to help communities organise themselves to reap extractive benefits of wild spaces. Maybe biologists would engage more in understanding landscape fractions as islands and stepping stones of biodiversity; and epidemiologists would research more on public health implications of these settings.

Perhaps this book is going to spark new dialogues on governance of nature and placemaking in cities; or perhaps become another unnecessary work on the shelf, criticised for its relevance, or even forgotten with time, ridden with isolation from people who are doing more important, more worthwhile inquiries on wild urban nature. But it goes without saying that the process of producing this work has allowed me to spend priceless time, and immerse myself, in an enquiry I deeply cherished. Thank you for picking up my book. Now let me share with you, a world I discovered.

PROLOGUE

The first sight of a cyborg assemblage

"Don't plants growing on just anything look aesthetic?", remarked my friend as we looked at a photograph of an old building covered with tree roots. This threw me back down memory lane, when I was having to push through the thick noisy crowd of traders, buyers, and passers-by in a busy market area, at the heart of Bengaluru in India. The aroma of fresh vegetables, flowers, and spices being sold here were temporarily interrupted by the odour of sweat from people's hard labour indulging themselves in the street economy. From locally-produced plastic toys to imported stationery; from furniture made of wood and steel to jewellery crafted from gold and diamonds; one could find almost everything here. Yet, the scene was far from perfect – although a part of the central business district, the Krishna Rajendra market (KR Market) wasn't a place with buildings with glass façades, high rise, and ornamental landscaping for civic beautification, typical to that of a global city. Quite the opposite: it was messy with conflicting movement of people on foot – sometimes accompanied by their pull-carts loaded with goods – who were all trying to find a way to get their business done. Motorised vehicles, including two wheeled scooters, three wheeled autorickshaws, and four wheeled sedans, that had also got caught in this crowd, honked out of boredom and regret of having chosen to pass through these streets. Despite being regularly cleaned, litter here inevitably piles up over the day, and stagnating water from leaks and overflows mixes with rotten vegetables to create mucky roadsides. At first sight, the scene was almost set in monochrome – the darkened soil from contamination and decomposing of organic matter; the black tar of narrow roads that twisted like serpents amid the compact layout of buildings; and weathered façades of buildings that had scraped their paints to turn grey.

Amid this chaotic scene, one could occasionally find cows or goats owned by nearby herders calmly taking advantage by feeding on the disposed vegetables, or people sipping tea in nonchalance, while taking a break from their hard day's work. As for me, I was making my way down the narrow streets of this old "pete area", to find a shop which sold handmade papers at a reduced price. *Pete* is a local word for town; and this market named *Chickpete* (meaning small town in Kannada) can be traced back as Bengaluru City's origin.[4] The market's 500-year-old legacy has become part of everyday life in Bengaluru – it is densely built and is devoid of greening efforts from city government or the land-owners. But amid the grey tone of worn-out paint and falling plaster I spotted the giant tree growing on top of a building roof. Thereafter, I continued to visit the market and found more places where plants had grown on their own, outside of human care. The market became my first study ground in search of spontaneous vegetation growing on buildings.

Bengaluru's KR Market lies in the innermost part of the city. It is also one of the oldest areas and thus significant in terms of historicity and culture. But this traditional setting is increasingly becoming obsolete as the current preferences of citizens are met with new developments in other parts of the city. Old city areas may still be thriving hubs of economy and housing but with narrow roads (that were not planned for the anthropometrics of automobiles) and overworked drainage systems that no longer match up to the requirements of high-intensity footfalls and activities which the city centre draws. Originating sometime in the mid-1500s, the layout of the *pete* area was not designed to accommodate high-rise buildings, and, over time, plots have gone through many subdivisions to turn into smaller, denser blocks. Several houses in the area also remain vacant and neglected – creating conditions for botanical growth.

Such occurrences of botanical growth were rare, but certainly worth the find. Each site was unique – some building walls had trees growing on them, typically stranglers (e.g., fig tree); a variety of ferns had grown around leaking pipes, peeping out of niches from the shafts. Small plants and grasses which barely found any space on the ground, grew in small patches on the rooftops. The sites varied in their architecture and the patterns of their drainage. It was intriguing to think about the different ways in which soils must have formed which allowed substrates for this biological growth.[5] With each site visit, I started to notice these assemblages more – what was earlier invisible in the city suddenly became replete. This notwithstanding, I also found that such assemblages are more likely to appear in certain parts of the city and not everywhere.

Let me elucidate. The context of the city centre was clearly distinct, as similar botanical growth was not necessarily common in other parts of the city – say for instance, unlike another locality in Bengaluru where I grew up, where there was no wild growth on buildings or on street-sides. It is worth noting the similarities and differences of the two neighbourhoods: first, although both the *pete* area and my residential locality were densely packed, the *pete* area was old and decaying; whereas the locality in which I lived was relatively new. Both areas were bustling with economies and people, and yet, homeowners in my locality seemed to have more incentive to maintain their properties. This is because in my locality, the prevalent economy was not trade but renting out of residential space. In the residential localities, most houses were meant for inhabiting – this means that homeowners had to maintain habitable conditions to attract competitive rental prices. Whereas in the KR Market's *pete* area – and as the name suggests – the main activity was trade, where residential houses were often converted to *godowns* (or warehouses) for storing goods to cater for the market. This meant that maintaining habitable conditions was not the primary need. Noticeably such conditions can lead to neglect of the built structure and allow plant growth to occur.

In major Indian cities, it is not at all uncommon for certain neighbourhoods not to have any green spaces – indicating, rather clearly, that the pace of globalisation has outrun the lives of people on ground. KR Market is not the only neighbourhood block in Bengaluru that was densely packed. The locality I grew up in, quite a distance away from this commercial district, also did not have any parks or playgrounds. I recall knocking on friends' doors and then marching down in a group to the playing fields in the adjacent neighbourhood. In the absence of parks and playgrounds, in my locality, residents had incorporated nature by adding planter boxes, or growing vines on rooftops in addition to appropriating small front yards for minimal gardening. This was by no means a norm, as most households did not keep plants. Besides, these settings of nature were rather predictable, and thus did not draw all my interest. Personally, I never kept plants myself – I knew they were a lot of work with the watering, turning soils, caring for the seedlings, and so on. Perhaps it was my laziness, or it could have been my affinity to the unfamiliar that desired to break away from the mundane and ordinary – in any case, my interest inclined towards the strange arrangements of wild spaces, that happened in places unintentionally, and were able to survive despite being uncared for. The meanings of these nature were unknown to me. Their mystery was infectious.

Notes

1 The accuracy of comparable data is hard to confirm as cities may calculate their green spaces through various approaches and at different points in time. Statistics from formal and grey sources on the Internet suggest the following figures for per capita green open space within the mentioned cities, over the last 5 years: Hong Kong (6.28 sqm); Shanghai (7.1 sqm), Mumbai (1.08 sqm), Hyderabad (2.5 sqm), Buenos Aires (2 sqm), Chihuahua (1.93 sqm), and Addis Ababa (1 sqm). These numbers are not prescriptive but indicative of the phenomenon of low urban green space.
2 Ample studies have been conducted to understand how this is disproportionately affecting the city's poor, who find it difficult to safeguard themselves from the harmful effects of pollution and climate change (Sarricolea, et al., 2022; Bhaviskar, 2018; Davoudi & Brooks, 2012). Research also suggests that the lack of green space also leads to deteriorating psychological and moral wellbeing – causing ailments such as anxiety and depression (see WHO, 2017).
3 See Konijnendijk, Ricard, Kenney, & Randrup, 2006.
4 See Deb, Dhindaw & King (2020), for a brief history of Bengaluru's spatial growth.
5 Although I often tend to use biological and botanical growth in an interchangeable manner, it is perhaps interesting to note how the term "biological" might refer to encompass plants as well as other microorganisms that might benefit from these sites as their birthplaces.

References

Bhaviskar, A. (2018). City limits: looking for environment and justice in the urban context. In S. Lele, E. S. Brondizio, J. Byrne, G. M. Mace, & J. Martinez-Alier (Eds.), *Rethinking Environmentalism: Linking Justice, Sustainability, and Diversity*. Cambridge, MA: MIT Press. doi:https://doi.org/10.7551/mitpress/11961.003.0009.

Davoudi, S., & Brooks, E. (2012). *Environmental Justice and the City: Full Report*. Newcastle University, Global Urban Research Unit, School of Architecture, Planning and Landscape. Retrieved February 7, 2023, from https://www.ncl.ac.uk/mediav8/apl/files/guru-ewp/ewp48.pdf.

Deb, A., Dhindaw, J., & King, R. (2020). Metropolitan Bangalore: crossing boundaries to integrate core and periphery. In S. Mehrotra, L. L. Lewis, M.

Orloff, & B. Olberding (Eds.), *Greater Than Parts: A Metropolitan Opportunity*. Washington, DC: World Bank.

Gelan, E., & Girma, Y. (2022). Urban green infrastructure accessibility for the achievement of SDG 11 in rapidly urbanizing cities of Ethiopia. *GeoJournal*, 87, 2883–2902. doi:https://doi.org/10.1007/s10708-021-10404-7.

Gerrish, E., & Watkins, S. L. (2018). The relationship between urban forests and income: a meta-analysis. *Landscape and Urban Planning*, 170, 293–308. doi: http://dx.doi.org/10.1016/j.landurbplan.2017.09.005.

Konijnendijk, C. C., Ricard, R. M., Kenney, A., & Randrup, T. B. (2006). Defining urban forestry – a comparative perspective of North America and Europe. *Urban Forestry & Urban Greening*, 4, 93–103. doi:https://doi.org/10.1016/j.ufug.2005.11.003.

Rigolon, A., Browning, M. H., Lee, K., & Shin, S. (2018). Access to urban green space in cities of the Global South: a systematic literature review. *Urban Sci.*, 2(3). doi:https://doi.org/10.3390/urbansci2030067.

Sarricolea, P., Smith, P., Romero-Aravena, H., Serrano-Notivoli, R., Fuentealba, M., & Meseguer-Ruiz, O. (2022). Socioeconomic inequalities and the surface heat island distribution in Santiago, Chile. *Science of The Total Environment*, 832, 155152. doi:https://doi.org/10.1016/j.scitotenv.2022.155152.

WHO. (2017). *Urban Green Space Interventions and Health: A Review of Impacts and Effectiveness*. Regional Office for Europe. Copenhagen: World Health Organization.

1

MORPHOLOGY

Small-scale nature in cities

City planning and design has tended to look at backyards, front yards, avenues, boulevards, and parks – which are very macro, very visible objects of nature. There are also various other typologies of urban nature that exist, which are sometimes referred to as urban wilderness; and have typically been studied as wetlands, city forests, vacant lots, cemeteries, railway verges, and so on (Gandy, 2022; Jorgensen & Keenan, 2011; Kowarik & Körner, 2005). These configurations also refer to relatively large tracts of land; as strips or blocks that can span from anywhere between a hundred square meters to several hectares. However, smaller quantities of nature emerge in cities as well, sometimes even invisible to the eye. These could be patches of grass, shrubs that bear wildflowers and berries; or single trees that grow on dilapidated buildings, disconnected from the earth; or ferns growing below the grate of a storm water drain. These configurations are very different from potted plants or wall mounters – simply because they have emerged unintended.[1] They may be described using an array of vocabulary including "accidental", "incidental", "spontaneous", and so on.

Some researchers have attempted to classify[2] spontaneous urban nature based on size of the patch they grow in (Rupprecht & Byrne, 2014b; Rupprecht & Byrne, 2014a). However, my approach is a bit more

DOI: 10.4324/9781003374602-1

flexible; because the idea of bigness and smallness in scale may not always be determined by a fixed measurement, and is almost always relative to an object's surroundings. Smallness in scale sometimes means that nature has grown in residual spaces, where land cannot be appropriated for anything else.[3] But plants may also grow spontaneously where landscape plans were drafted, but never really realised due to high maintenance cost or negligence of the agencies responsible for their upkeep. It is unlikely, that people would make a fuss about the existence (or loss) of a small flowering shrub or a patch of grasses along the road; which might measure up to say, only a size of a shoe, or a refrigerator, or a minivan. But they exist nevertheless, and add to urban nature. Such nature is also more easily lost – sometimes as a result of repair works, or other times due to conscious drives of "de-weeding" or "paving" involved in landscape maintenance regimes. Unintended, spontaneous, or informal natures are powerless, because they lack legitimacy for a variety of reasons that may relate to use of land parcels, types of landownerships,[4] low utilitarian values, risks posed, or simply a lack of visual appeal (more about these in Chapter 3 on Existence).

Despite small-scale nature's lack of legitimacy, they should still fall under the broad umbrella of urban forest; as the urban forest is a composition of all trees, plants, and shrubs in cities (see Miller 1997 cited in Konijnendijk, et al., 2006).[5] Yet, the role of small-scale nature has not been adequately explored. Given how parks and playgrounds are islanded within cities amid large tracts of dense concrete and tar, it would seem as though large green spaces with lots of plants compensate for the lack of nature in between. That is to say: in comparison to the role of large green spaces, small-scale natures do not appear significant.[6] Then it was perhaps hard for the residents to imagine the consequences of deciding to uproot and clear-out a small patch of wild plants that had grown along their neighbourhood street. However, in the book, *Order of Things*, Michel Foucault compares the writings of Buffon and Charles Bonnet to point out how both of the eighteenth-century naturalists share the same concerns of nature not having leaps (Foucault, 2005). This imagination of nature as a continuum then resonates and if, indeed, there are no leaps in nature, it begs the question:

- What is the role of small-scale spontaneous nature in cities?
- By removing small-scale assemblages of spontaneous nature, are we unknowingly disrupting critical processes within the urban forest?

To answer these questions, this book draws insights from first-hand documentation of small-scale nature in cities; particularly Bengaluru and Kolkata (in India), Whitby (in the United Kingdom), and Quito (in Ecuador). By doing so, the book uncovers lives of, and lives around, unique settings of plant growth that have come up in the oddest places within cities. This book refers to such arrangements or settings of small-scale nature and built-environment as "assemblages".

Organisation and forms of assemblages

In urban areas, sometimes, nature grows where we least expect. Plants grow not only on buildings but also in residual spaces – fused with human-made elements such as automobiles, mechanical water pumps, and electric transformers. To not notice these small-scale assemblages made of nature and human-made objects, would be to ignore a fundamental unit in the production of urban nature – thereby urban space itself. Within urban limits where land is largely paved or built upon, nature will have to find its way from amid human-made elements. Thus, trees or other vegetation growing on buildings can rarely be viewed separately from human-built surfaces or objects. The human eye captures everything in and around the biological growth, and when it finds an assemblage worth noticing, it has found a "Landscape Fraction".

Landscape fractions can be made of wild plants growing on buildings or in interstitial urban spaces. Such assemblages are spontaneous and were not intended to be in their existing setting – neither by humans nor by nature, alone. Though this definition illustrates more of what such assemblages are not, the description serves two purposes: one is to identify, distinguish, and confirm the presence of both the "natural" and the "built" involved in the composition of landscape fractions; and another is an attempt to fuse the two (natural and built) while avoiding an immediate urge of constructing a hierarchy between them.

Landscape fractions can also be designed assemblages. Small-scale assemblages of nature are used as greening tools by architects and urban designers. Rooftop gardens – creating an appeal of "meadows in the sky" (Loder, 2020, p. 104) – are not uncommon in hotels, offices, and residences; serving as great spaces for relaxation and social gatherings. Sometimes public spaces that do not have a lot of planting space are revitalised by green walls to add to the green aesthetics of the city. Such assemblages – i.e., wall and roof treatments with plants – are also retrofitted in architectural designs to make buildings reduce energy consumption and thereby making them more sustainable. In Mexico, pillars on the highway that support elevated roads have been retrofitted with vertical planters, so as to appear green (Jenkins, 2018). In Bengaluru, the same has been experimented with, but has not always been successful. Indeed, such landscapes need care and suitable environmental conditions to thrive.

On the other hand, and rather effortlessly, assemblages of spontaneous nature can be seen to emulate the aforementioned designs of landscape architects by growing on built spaces abandoned by humans. Reportedly, in Noida (India) about a hundred residential buildings now lie vacant, that have now created a "post-apocalyptic" appeal (Anand, 2019). The newspaper article shows photographs of buildings in their bare concrete frames and roofs, to capture how constructions have been halted midway, allowing plants to grow. The unfinished concrete structures are more prone to absorbing water and thereby fostering conditions needed for plant growth.

Formations of assemblages have a close relationship with the spatial organisation of the city. Spontaneous nature grows almost everywhere in cities if left untouched by human action or where there is residual space. Yet even amid the ubiquitous existence of natural growth in cities, certain areas of the cities are particularly good breeding grounds for assemblages that are formed by a group or individual plant species in symbiosis with human-made elements. Old city areas typically have traditional architectural heritage which have undergone natural withering to create conditions for plant growth. Other highly probable microregions within the city are low-income neighbourhoods where homeowners are unable to pay for repairs, and local governance is fund-deficit which leads to neglect of designed landscaping in public spaces (e.g., roadside verges or public parks). Many cities have gone through or are also going through de-industrialisation, which means that formerly operational industrial yards are now decommissioned or locked-up campuses; awaiting their redevelopment. Meanwhile, nature has found a safe haven in these areas.

Small-scale assemblages of spontaneous nature, even while abundant, are like tiny dots on the city canvas. There is also very little literature that help us understand such ensembles of built and natural. So far, what Rupprecht and Byrne (2014b) refer to as "microsites" have been underrepresented in our conceptual understanding of natural networks.

Perhaps superimposing the patch-corridor-matrix of landscape-ecologist Richard T.T. Forman helps in accounting for both the fragmentation and size of ecosystems in urban regions. Small patches which Forman describes as "fragments of nature" have been assigned the lowest priority albeit acknowledging that they could offer benefits such as acting as stepping stones for species movement (which is but a less-efficient and discontinuous version of corridors) (Forman, 2008, pp. 38–40). Notably, within Forman's hierarchy of small patches, the smallest units discussed are city parks. And smaller than that, is our dotted existence of plants growing out of residual urban spaces as well as unexpected habitats created on walls, paving, and rooftops. Thus, to offer a resolution that adequately captures the existence of small-scale nature and microsites, one can perhaps add another class to Forman's framework to represent it as: 'dots-patch-corridor-matrix'.

Landscapes within landscapes

Landscape fractions could also be just an aesthetic figment of an otherwise regular unintended landscape. When a park is the landscape, and the wild growth is a patch within – the inner landscape is the fraction where natural growth includes a variety of colourful wildflowers – such as *Lantana camara*, an invasive species. Butterflies are attracted to them. So are bees. Being pollinators, bees and butterflies are essential for the health of natural ecosystems in the city.

In Bengaluru's peripheries, a park lies neglected by the municipal authorities. A few giant trees that create shade within were pre-existing before the area was fenced off to demarcate the public park. At least a decade ago, the park was established with good infrastructure (kerb stones, seating, and walking track). However, over time the seating has corroded and there has been little or no maintenance. For a long time, the park has had only one lamp post with a single halogen bulb for night time activities; that is operated by people who visit the park. This notwithstanding, street lights outside the park and light from surrounding buildings offer some amount of luminance for a stroll in the night. Still there would be patches of darkness if just one of these light sources were turned off. The park is located in a low density and relatively disconnected neighbourhood on the outskirts of the city, where schools and shops only cater to the neighbourhood population.

There were more trees within the park, which have now been removed, because they were health hazards – no new trees have been planted, though there are planter rings which indicate that this was planned when the park was conceived. Small ornamental plants were planted, but barely a few survived due to a lack of continuous care. Yet, the lack of designed landscaping does not stop people from using this area. This park is in fact one of the most vibrant spaces of the community where religious, educational, and recreational activities (such as jogging, children's assembly for an adjacent public school, martial arts classes, and other social gatherings) are held. The central open space is used for a variety of sports: sometimes children set up wickets to play cricket, other times a couple of goalposts for football, or even tie a net across two poles for badminton.

For the most part, the wild growth is not seen as undesirable by users; but it does take away from the play area of the park according to some: "If the plants weren't there, we would have more space to play. It is a good thing, if [authorities] could remove the [plants, fallen branches, etc.]", said a user who would be a high school student.[7]

Interaction with nature is not always planned yet can be pleasurable – at least for the moment – sometimes for the viewer, and other times for those interacting physically from closer quarters. For instance, when the cricket ball is lost in the wild bush, and the kids go in search of it, the interaction with the bush is purely incidental. They would interact with a Parthenium bush and the landscape fraction that is formed by a sea of this exotic plant within a designed arena – the park. This wild bush has in fact invaded a large portion of their play area due to the neglect of the park authorities, albeit remaining neatly bounded within the jogging tracks.

While the photographer viewing from afar enjoys the green visual, the children searching for the ball are faced with a perplexing problem that they must solve, i.e., finding the ball. At the end of this problem is a reward – the ball itself. The boys would use sticks to beat the grass to scare off snakes and other animals which might cause harm. Other times they would swing their cricket bats to trim the thick bush so as to see clearer, and to help find the object they are looking for. This of course makes little difference to the landscape as the invasive Parthenium bushes are known to grow rapidly. But for the children, it is nevertheless a tiring activity which can be disappointing if the ball is not found.

Meanwhile, the viewer from afar is deprived of the smaller details of this setting, which become part of the children's experiences and memories. There is a metal water pipe sticking out of the bush; it can only be seen from up close as it otherwise becomes hidden within the plants. The landscape is not just comprised of the Parthenium bush, which is visually dominant, but of several other plants which house interesting and colourful creatures within itself (see Figure 1.1). There are the miniature bouquets of wildflowers on one hand, and hidden lives of butterflies and caterpillars on the other. One can only see this up close. Thus, the charm of this landscape is different from afar and near; and especially in being composed of miniature landscapes within the larger landscapes.

Landscape fractions can also be formed by combining elements that are natural and human-made. In the playground site explained above, there was only one human-made element – the water pipe. On the other hand, in some of the landscapes, the human-built elements are more dominant. Vegetation sometimes grows around machines or "technological relics" such as electricity transformers and control boxes for water supply placed alongside the road. From afar, the mechanical devices appear as unusual elements cropping up amid grasses and shrubs, often partly hidden in them.

FIGURE 1.1 Wild growth in a neglected space: children search for a missing cricket ball amid a patch of flowering bushes within a neighbourhood park of Bengaluru. October 20, 2019.

To illustrate the composition of technological relics and nature, another site I explore is a control box within a residential area of Bengaluru that has been placed at the edge of a road (Figure 1.2). The equipment installed by the local government is connected to the public borewells on site, that supply piped water to households in the area. Control boxes are often a retrofit or built after the area has been developed – that is an obstruction on the road for vehicles and pedestrians. But this unplanned obstruction (often thought of as negative spaces in design) also creates residual spaces for nature to grow. With no definite schedule, but perhaps three times a week, the public worker assigned to release water to the neighbourhood households uses his long device to reach underground through a cavity to open the valve located right beneath the control box. The joints near the valve could well be susceptible to leaks; providing a water base to create conditions for vegetation growth. The cavity or hole in the ground might offer refuge for small animals, hidden from human eyesight due to the plant growth.

FIGURE 1.2 Associated with technological relic: wild plants growing around a control box for water supply along a neighbourhood street of Bengaluru. October 20, 2019.

These incidental settings could well become laboratories for studying micro-ecosystems.

Control boxes placed by Bengaluru city's water board are a common sight along roadsides in the peri-urban neighbourhoods.[8] Although installed for public welfare by the government, these devices are seldom accessed by neighbourhood residents themselves. Specialists who are employed by the city authority come regularly to turn-on and turn-off the water supply, and at times to fix it i.e., if the device starts malfunctioning. So, humans rarely come near this device, and more so, only those humans who have specialised skills. But in this rare interaction limited by time, engineers or mechanics are able to look up close into the micro aspects of nature around a box. They come in close view of insects, amphibians, and reptiles: lizards, frogs, ants, butterflies are a common sight. Not to mention the fact that the miniature flowers which bloom amid the grasses are a pleasant sight. Combined by human and natural elements, such landscapes offer patches that are intense micro habitats of plant and animal diversity. It is not always safe as snakes are found to enter and coil inside such boxes. But where else are the city-dwelling children going to find opportunities to explore nature and learn about the dos and don'ts of interacting with machines in nature?

Mapping biotopes of urban-nature

Spontaneous nature can be found growing on building walls, around machines and other instalments of a factory, on roofs and chejjas of a residential building, in the alleys, along streets, along railway tracks, and so on. They can emerge just about anywhere; which is why these formations are also unpredictable, and difficult to document. Satellite imagery might not map flora that is particularly fused with buildings because plants grow on vertical walls too, and the extent of surface area they cover can't always be captured by aerial imaging. Sometimes plant growth is also hidden under structures like open sheds or under suspended building connectors and grates of a street-side drain. Thus, ground truthing becomes necessary. As one noteworthy experiment, botanists in Berlin and Yokohama have been able to map biotopes across the cities through on-ground site surveys Muller & Fujiwara, 1998).

There have been long-standing efforts in parts of the world to understand arrangements of nature typical to urban areas, but not many studies that reveal the arrangements that combine both built and natural.

Urban administrative areas such as the city of Berlin and Tokyo-Yokohama urban agglomeration practice classification and mapping of biotopes (or typical natural habitats). Some of these sites have been found to support endangered animal and botanical species such as bats in Yokohama and bluegrass in Germany (Muller, 1997; Gandy, 2017). However, there was not enough detail within the mapped biotopes of classifications that would recognise plants growing on built spaces. Mapping of recognised biotopes of Berlin had 11 classes as of 2013 covering a total of 5016.67 hectares of legally protected sites, of which only 2.64 hectares indicated an assemblage of vegetation built-up over buildings and traffic facilities.[9] This signifies the rarity of human-nature assemblages and sparks an opportunity for research in cities like Bengaluru and Kolkata which have several such sites.

Informal urban green spaces are often a composition of wild grasses, shrubs which are typically invasive, ivy, and strangler trees – all of whom can thrive under harsh conditions such as scarcity of land, hard medium for penetration of roots, or depleting groundwater table. Stranglers such as banyan and fig have successfully adapted to the urban built conditions of Kolkata as well as Bengaluru. The ability of their roots to penetrate through pores in the masonry of buildings is an advantage. Though there are a variety of ferns and ivy that grow on the walls, early succession plant species –such as Doab, Lemon, and Elephant grass (*Cynodon dactylon, Cymbopogon citratus*, and *Pennisetum purpureum Schumach*) as well as shrubs such as Parthenium (*Parthenium hysterophorus*) and Lantana (*Lantana camara*) – are not uncommon on ground surfaces such as roadsides and walkways. In Bengaluru and Kolkata, I also found beds of Parthenium and a variety of tall grasses growing on flat rooftops which remain locked down – recreating formations of nature on the terrace to appear like roof gardens.

An understudied aspect of urban nature is flora on the vertical surfaces. Yet a variety of fern species have adapted to masonry walls and thus such ensembles are a common sight around the world. Acknowledging this, Reis, Lombardi, and Figueiredo (2005) explain masonry walls as: "specialized microenvironments conditioned by human beings". Somewhere near the Hedua area of Kolkata, I was able to spot an invasion of ferns on the boundary walls of what seemed to be a small-sized factory establishment (see Figure 1.3). Hedua emerged among the early colonial areas of Kolkata and today is largely a residential area. The assemblage of ferns and red brick masonry along with colonial-style piers are reminiscent of historicity.

FIGURE 1.3 Wild nature on vertical surfaces: ferns growing on exposed masonry walls of a small-scale factory in Kolkata. July 14, 2018.

The plants had created a visual repose in the alley which is otherwise devoid of much vegetation. Notably, while many other walls in the traditional core of Kolkata bear biological growth, this happens to be a site where I found the ferns to be more successful in dominating space among stranglers and other floral species typically found growing on masonry walls. However, my admiration for the natural ensemble as a one-time visitor with a purpose of writing a book on it, is not something another everyday passer-by might hold.

Small-scale nature growing on walls is often ignored or gets discounted due to their perception of being transitory. During my walks around the city, inspecting plants growing out of built structures, people have often shown curiosity towards my interest in these assemblages. Visiting one of the streets in Kolkata, an exchange went something like this:

> "What are you doing?"
>
> "I am taking a picture of these", I pointed to the fern growth on the walls of the old factory unit.

> "Are you from here? Are you looking to buy the property? Would you like me to contact the owner?" the man, a passer-by, went on as if the plant growth wasn't even there. Nature was invisible.

Often when I explain that I am taking pictures of the spontaneous vegetation, people respond by saying "they grow from time to time".[10] People are also being apologetic of not having removed them when they say, "we do cut them down, yet they grow back.".[11]

Mapping microsites of spontaneous nature can help quantify their existence, and therefore helping measure roles they play more accurately. Though such a fine-grain and resource intensive exercise at the city-level is beyond this book's scope, some broad patterns at the urban scale can safely be assumed i.e., assemblages of minimal nature and built are more likely to occur in sites of decay and dereliction – or sites of age; biotopes in an industrial parcel would differ significantly in character to residential neighbourhoods and so on.

Reversing globalisation: building a city bottom-up

Cities are hotspots of plant biodiversity. Connected with global markets, cities also import a variety of exotic flora in parks, offices, and homes. A wide variety of substrates are available from near and far which offer a range of growth media for plants. Nurseries are able to offer care during the early phases of the plant growth and bring to the market colourful, vibrant, fragrant, and rare plant species. This has been possible because of a high transmission of knowledge and skills along with transportation of plants from and between different parts of the world. The city has become the node of creation of novel landscapes; where landscape architects, biologists, and other plant-care professionals come together to make marvels that leave people in awe. These are then photographed and distributed through print magazines, the Internet, and television around the world; and in turn influence global aspirations of wanting to be in these high-concept designer environments.

However, global aspirations are far removed from the lives of the majority of urban populations; and it is rather evident that such aspirations have completely failed for certain sections of the population who are unable to see nature in their immediate surroundings; let alone landscape designs that are more sophisticated. A globalisation of aesthetics, that is propagated through magazines and other popular media, looks at imports of flora from various parts of the world. Real estate developers are able to transact high-end purchase deals through such designer landscapes. To a great extent, this flow of material and ideas attracts an immense amount of diversity in urban landscaping which can create built-environments that are exotic, intriguing, and exciting. Having said that, such configurations are not always affordable.[12] They require skills, knowledge, money, and time which are more readily available to the affluent population. To address the resource gap, both the government and people choose to pave public spaces and seal the exposure of soil from where plants can grow.[13] This is not solving the main problem i.e., a deficit of green cover in the neighbourhoods.

Global forces don't just shape landscapes, but also set expectations of what cities should be like in the future. Planning is a complex process that has traditionally involved a top-down bureaucratic and political mode of envisioning change. More recently, there has been a growing participation from the civil society and private sector. However, town planning still remains a conception of what cities aspire to be. The gap between what is envisioned and what is achieved is huge. Lack of funds, and sometimes socio-political disruptions almost always push cities to undertake mid-course corrections. Other times, cities have to accept a harsh reality i.e., the plans and designs are not always realised.[14] These processes of production of space relate to Henri Lefebvre's notions of perceived versus conceived space (Lefebvre, 1991).[15] While on one hand, several cities – particularly in the Global South – struggle to meet the basic WHO prescription of a nine sqm green space per capita (Gelan & Girma, 2022), on the other, there is evidence to suggest strong inequalities in distribution of urban forest cover within cities in the developed as well as developing countries (Gerrish & Watkins, 2018; Rigolon, Browning, Lee, & Shin, 2018). For the areas within cities, where the pace of globalisation has not matched up with people's lives on the ground, city makers can opt to rethink cities bottom-up.

In reality, however, nature seems to have its way to counter the forces of its expulsion through globalising forces; and assemblages are more than just dots upon the urban landscape. In absence of human care, nature not only has to compete for suitable "land", but also resources. Upon viewing closer, one can find the roots of the fig tree (*Ficus religiosa*) growing on a wall, to have penetrated through the little pores, deep into the mortar joints; the grass to have grown upon the roof from a substrate deposited as soil, creating a habitat for insects; and vines to have crept up street-side electric poles by intertwining around wires that rise up to the buildings for supplying the digital services of cable TV. Together they construct a visual that is only part nature because the plants grow on, or around, human-built objects. Sometimes, the built structure can be as simple as a footpath from whose cracks emerge certain species of grasses; or kerbstones of a jogging track along which a few wild shrubs have grown. Notably, the type of species that grow would depend on a variety of conditions which are both local and regional. Local factors include substrates for growth, the nature of nutrient base available, and supporting structures as required for specific species to grow. Regional factors would not only be related to socio-economic processes such as migration and income but also with respect to plant communities belonging to particular biomes within which the cities are situated (Roman, et al., 2018).

Given aggressive human attitudes towards non-designed nature, rendering them as "weeds" and "signs of dereliction and abandonment", plants find very little space to successfully thrive within cities. We are dismissive towards such nature. This perhaps relates to Foucault's idea of *episteme* that refers to a subconscious set of beliefs guiding humans to what they should believe and what they should refute (Foucault, 2005). Later, and particularly in Chapters 2 and 7 (on Meanings and Frames), I will explore what some of the mental constructs surrounding spontaneous nature are, and how they are linked to the spatial, political, and socio-economic context of landscape fractions.

Spaces of wild growth in cities can hold a dual character of positive and negative perceptions in people's minds. In 2007, Anna Jorgensen and Marian Tylecote's work explored this as "ambivalence" of nature (Jorgensen & Tylecote, 2007). The authors make the case for valuing urban wild spaces in cities; as such landscapes, despite their docile character compared to natural wilderness, come closest to offering an experience of wilderness to the urbanites who are otherwise growing increasingly detached from the naturalness of the countryside. Geographer and urbanist, Matthew Gandy, notes that natural regenerative vegetation in marginal spaces within the city otherwise referred to as "wastelands" or "terrain vague" – broadly encompassing derelict or abandoned areas – can hold values of aesthetics and offer spaces of culture, creativity, and imagination (Gandy, 2013; Gandy, 2016). While these concepts have been developed from reference to relatively large parcels of nature, this book investigates their transferability for fractions of nature within cities that are far smaller in scale.

Studying fractions of urban nature, whose representative scale can be described as buildings wrapped up with plants, can become a myopic exercise. One would typically have to somehow weave insights from studying small-scale spontaneous nature back to gain insights at the city level. Urban ecologist and botanist, Peter Del Tredici in his article *Spontaneous Urban Vegetation: Reflections of Change in a Globalized World*, discusses spontaneous nature as a microcosm of economic processes at the city level (Del Tredici, 2010). I will elaborate some more upon these in the next chapter on Meanings; but broadly, the study of landscape fractions in this book should present a greater resolution to Del Tredici's existing notions (of economic failure and success) affiliated with spontaneous nature. The fairly ubiquitous presence of such nature provides the basis to suggest that small-scale settings of wild biological growth are an integral part of everyday nature for the urban population.

Cities are competing with each other to become global brands through iconoclastic architecture, market specialisations, and vibrant lifestyles; such has been key to raise the image of a city and draw talented labour and businesses.[16] Scaling-up seems to be the aspiration of every business and advertisements feature extravagant lifestyles on TV, on product packaging, and on billboards that send subtle cues suggesting: bigger is better.[17] Large parks and waterfront developments, vibrant night life in bars and pubs, bigger malls and amusement parks, taller residential towers and office buildings, and wider roads are the images that cities convey to demonstrate financial power and quality of life. The imaginaries of a better quality of life have led to a rapid pace of urbanisation which, evidently, cities around the world are struggling to keep up with – for it is common to find high-density environments in pockets that have been built compromising even the basic requirements of green spaces. As a measure of damage control, Jennifer Wolch, Jason Byrne, and Joshua Newell have stressed on the importance of small-scale nature in cities to deliver ecosystem services (Wolch, Byrne, & Newell, 2014). The authors advocate that the potential of small-scale green spaces should not be ruled out just because they do not offer the same ecosystem services as that of the large green spaces; and, instead, should be valued for whatever services they can offer.

Urban green spaces are an opportunity to help people reconnect with nature. Timothy Beatley's long-standing body of work has pointed out that in urban areas where people often have less physical interaction with natural landscapes, education in urban-natural settings can help people to develop an emotional connection with nature (Beatley, 2017; Beatley, 2011). Other studies have widened this understanding by confirming that people with exposure to vegetation in their surroundings develop "pro-environmental attitudes" or are more inclined to stand up for environmental causes like climate change which are not necessarily a local phenomenon (Whitburn, Linklater, & Milfont, 2018). This connect between people and nature is particularly crucial as Dobson and Dempsey (2020) argue as to how the focus on global environmental crises has, at the same time, led to ignorance of local problems such as degradation of space which can enhance the everyday life of the citizens. This notwithstanding, while a strong link has been established between local nature and global environmental responsibility, it is unclear as to what extent micro- or small-scale nature can help urbanites reconcile with nature which is gigantic in scale.

Spontaneity of nature already exists in cities, and can't be created; for if it is created, then it is not spontaneous. In theory, the services of wild spaces may be replaced and made more efficient through design. Though this perception does not reflect the universal realities of cities – especially where economic (e.g., lack of resources to replace wild growth with designed landscape) and structural limitations (e.g., when trees growing on buildings cannot be separated without causing severe damage) make it less logical to remove spontaneous growth. Thus, it warrants the commencement of a search for models of landscape management and design grounded in local realities. Such models of landscape management are needed, particularly in cities of developing countries where poverty and inequality are dominant issues; health and wellbeing are compromised; austerity is on the rise; and biodiversity is threatened. However, as French anthropologist, Bruno Latour in his book *Down to Earth: Politics in the New Climatic Regime* notes, any efforts to move away from the global scene is now labelled as a primitivistic approach; and that we are stuck in a space and time where we are neither able to deliver the global promises nor encourage local visions (Latour, 2018). Latour thus suggests the search for a middle ground, where alliances can be possible.

Notes

1 Matthew Gandy conceptualises unintentional nature as "an aesthetic encounter with nature that has not been purposively created" (Gandy, 2016). More on this is discussed in Chapter 4 on Aesthetics.

2 Studying two cities, Brisbane (Australia) and Sapporo (Japan), Christoph D. D. Rupprecht and Jason A. Byrne arrived at a nine-fold typology of informal green spaces (IGS). The classes identified were street verges, lots, gaps (e.g., between walls and fences), railways (e.g., along the tracks), brownfields (e.g., factory grounds and industrial parks), waterside (e.g., edges of rivers and canals), structural (e.g., walls and roofs), microsites (e.g., cracks and holes), and electrical lines (e.g., space below powerlines). The authors categorised these based on small, medium, and large sizes where microsites, street-side verges, and structural sites are labelled as small. Small size IGS were found to vary anywhere between less than a square meter to 100 sqm (Rupprecht & Byrne, 2014b; Rupprecht & Byrne, 2014a).

3 K. Keswani. Appropriation of small green spaces, interview with A. Deb, December 4, 2021.

4 K. Keswani. Appropriation of small green spaces, interview with A. Deb, December 4, 2021.

5 One of the earliest definitions of urban forest refers to Robert Miller's book "*Urban Forestry: Planning and Managing Urban Greenspaces*" that describes the object as "the sum of all woody and associated vegetation in and around dense human settlements, ranging from small communities in rural settings to metropolitan areas" (see Miller, 1997, p. 27 cited in Konijnendijk, Ricard, Kenney, & Randrup, 2006). To add to this, Konijnendijk points to how the concept helps perceive the otherwise "individual greenspace elements as part of an integral whole" (see Konijnendijk, Ricard, Kenney, & Randrup, 2006).

6 More recent literature has also begun referring to urban forest as "all the trees", omitting the role of shrubs and other ground cover.

7 Informal field discussion with park user, 2021, Bengaluru, India.

8 While the centre of Bengaluru is supplied with piped water drawn from river sources far away, in parts of the urban periphery, government service providers draw groundwater, to distribute it to households via pipes. Typically, one borewell is allocated for, say, every three streets.

9 Senate Department for Urban Development, Building and Housing. Berlin Environmental Atlas: Biotope Types 2013 https://www.berlin.de/umweltatlas/en/biotopes/biotope-types/continually-updated/map-description/; For more, see also: https://www.berlin.de/umweltatlas/en/biotopes/.

10 Incidental interaction with passer-by, 2018, Kolkata, India.

11 R. Sircar. Managing biological growth on a wall on Amherst Street, interview with A. Deb, July 16, 2018.

12 E. Longchar. Attitudes towards botanical growth, interview with A. Deb, January 2, 2022.

13 See Figure 6.3 in Chapter 6 on Coproduction for before and after concreting.

14 B. Banerjee. Wild growth and urban planning, interview with A. Deb, December 9, 2021.

15 This refers to two components of the spatial triad given by Henri Lefebvre in his work *The Production of Space*. The perceived space is what is imagined through spatial practices such as planning, and conceived spaces are finally what is constructed in material space. See: Lefebvre, 1991. *The Production of Space*. III. Translated by Donald Nicholson-Smith. Cambridge, MA: Basil Blackwell.

16 This interconnectedness of cities at an international scale is explained well by Saskia Sassen's proposed framework of "Global Cities" (Sassen, 2005). However, Sassen's framework is also a high-level view; which accounts for cities that are beginning, increasingly, to have their influence across national borders. Yet in the same cities, a more significant force of urbanisation can be local migration from rural areas and other "secondary" cities that are not directly part of the global network of cities.

17 For example, shops display posters of "It's your life, make it large" an advertisement slogan of a popular whisky brand.

References

Anand, A. (2019, September 15). Ghost city: over 100 residential towers abandoned by builders in Noida. Retrieved June 8, 2022, from *India Today*: https://www.indiatoday.in/mail-today/story/noida-s-ghost-towns-more-than-100-residential-areas-abandoned-by-builders-1599265-2019-09-15.

Beatley, T. (2011). *Biophilic Cities: Integrating Nature into Urban Design and Planning*. Washington, DC: Island Press.

Beatley, T. (2017). Biophilic cities and healthy societies. *Urban Planning*, 2(4), 1–4. doi:10.17645/up.v2i4.1054.

Del Tredici, P. (2010). Spontaneous urban vegetation: reflections of change in a globalized world. *Nature and Culture*, 5(3), 299–315.

Dobson, J., & Dempsey, N. (2020). Why practice doesn't make perfect. In N. Dempsey, & J. Dobson (Eds.), *Naturally Challenged: Contested Perceptions and Practices in Urban Green Spaces* (pp. 1–7). Cham, Switzerland: Springer.

Forman, R. T. T. (2008). *Urban Regions: Ecology and Planning Beyond the City*. New York: Cambridge University Press.

Foucault, M. (2005). *Order of Things* (Reprint ed.). New York: Routledge.

Gandy, M. (2013). Marginalia: aesthetics, ecology, and urban wastelands. *Annals of the Association of American Geographers*, 103(6), 1301–1316.

Gandy, M. (2016). Unintentional landscapes. *Landscape Research*, 41(4), 433–440. doi:10.1080/01426397.2016.1156069.

Gandy, M. (Director). (2017). *Natura Urbana: The Brachen of Berlin* [Motion Picture].

Gandy, M. (2022). *Natura Urbana: Ecological Constellations in Urban Space*. Cambridge MA: MIT Press.

Gelan, E., & Girma, Y. (2022). Urban green infrastructure accessibility for the achievement of SDG 11 in rapidly urbanizing cities of Ethiopia. *GeoJournal*, 87, 2883–2902. doi:https://doi.org/10.1007/s10708-021-10404-7.

Gerrish, E., & Watkins, S. L. (2018). The relationship between urban forests and income: a meta-analysis. *Landscape and Urban Planning*, 170, 293–308. doi: http://dx.doi.org/10.1016/j.landurbplan.2017.09.005.

Jenkins, L. M. (2018, October). *Mexico City's vertical gardens: seeds of change or cynical greenwashing?* Retrieved April 7, 2022, from *The Guardian*: https://

www.theguardian.com/cities/2018/oct/30/mexico-city-via-verde-vertical-gardens-pollution-climate-change.

Jorgensen, A., & Keenan, R. (Eds.). (2011). *Urban Wildscapes* (1st ed.). New York: Routledge.

Jorgensen, A., & Tylecote, M. (2007, August). Ambivalent landscapes – wilderness in the urban interstices. *Landscape Research*, 32(4), 443–462.

Konijnendijk, C. C., Ricard, R. M., Kenney, A., & Randrup, T. B. (2006). Defining urban forestry – a comparative perspective of North America and Europe. *Urban Forestry & Urban Greening*, 4, 93–103. doi:https://doi.org/10.1016/j.ufug.2005.11.003.

Kowarik, I., & Körner, S. (Eds.). (2005). *Wild Urban Woodlands: New Perspectives for Urban Forestry*. New York:Springer Berlin Heidelberg.

Latour, B. (2018). *Down to Earth: Politics in a New Climatic Regime* (English Edition ed.). (C. Porter, Trans.) Cambridge, UK: Polity Press.

Lefebvre, H. (1991). *The Production of Space* (III ed.). (D. Nicholson-Smith, Trans.) Cambridge, MA: Basil Blackwell.

Loder, A. (2020). *Small-scale Urban Greening: Creating Places of Health, Creativity, and Ecological Sustainability*. New York: Routledge.

Muller, N. (1997). Biotope mapping and nature conservation in cities – Part 1: background and methods as basis for a pilot study in the urban agglomeration of Tokyo (Yokohama City). *Bull. Inst. Environ. Sci. Technol. Yokohama Natn. Univ*, 23, 47–62.

Muller, N., & Fujiwara, K. (1998). Biotope mapping and nature conservation in cities. Part 2: results of pilot study in the urban agglomeration of Tokyo (Yokohama City). *Bull. Inst. Environ. Sci. Technol. Yokohama Natn. Univ.*, 24, 97–119.

Reis, V. A., Lombardi, J. A., & Figueiredo, R. A. (2005). Diversity of vascular plants growing on walls of a Brazilian city. *Urban Ecosystems*, 9(1), 39–43. doi:http://dx.doi.org/10.1007/s11252-006-5528-1.

Rigolon, A., Browning, M. H., Lee, K., & Shin, S. (2018). Access to urban green space in cities of the Global South: a systematic literature review. *Urban Sci.*, 2(3). doi:https://doi.org/10.3390/urbansci2030067.

Roman, L. A., Pearsall, H., Eisenman, T. S., Conway, T. M., Fahey, R. T., Landry, S., … Jenerette, G. D. (2018). Human and biophysical legacies shape contemporary urban forests: a literature synthesis. *Urban Forestry & Urban Greening*, 31, 157–158.

Rupprecht, C. D., & Byrne, J. A. (2014a). Informal urban green-space: comparison of quantity and characteristics in Brisbane, Australia and Sapporo, Japan. *PLoS ONE*, e99784.

Rupprecht, C. D., & Byrne, J. A. (2014b). Informal urban greenspace: a typology and trilingual systematic review of its role for urban residents and trends in the literature. *Urban Forestry & Urban Greening*, 13, 597–611.

Sassen, S. (2005). The global city: introducing a concept. *The Brown Journal of World Affairs*, 11(2), 27–43.

Whitburn, J., Linklater, W. L., & Milfont, T. L. (2018). Exposure to urban nature and tree planting are related to pro-environmental behavior via connection to nature, the use of nature for psychological restoration, and environmental attitudes. *Environment and Behavior*, 57(7), 1–24.

Wolch, J. R., Byrne, J., & Newell, J. P. (2014). Urban green space, public health, and environmental justice: the challenge of making cities 'just green enough'. *Landscape and Urban Planning*, 125, 234–244.

2

MEANINGS

Era and age of city

"It's just moss. They are everywhere". My housemate uttered confused, when he found me entering the Endcliffe residence of Sheffield University, with a handful of moss. On a good Saturday, I had gone about my neighbourhood collecting these green lumps of micro growth, as I like to put them in jars to create moss terrariums. Moss is not uncommon in India, but certainly not as abundant as they are in the UK. The cold moist climate seemed perfect for moss to grow all year round; over walls, pavements, and building rooftops. Endcliffe – a neighbourhood located in the south-western part of Sheffield city – has a Victorian charm to it, and one can look at moss growth as something that adds to the age of the buildings. Indeed, local context matters, because in Endcliffe, virtually every house, every roof, and every boundary wall had moss growing on it. Moss growth was quite ubiquitous. In similar climates, "moss bombs" are quite popular; where graffiti is drawn using a concoction of moss mixed with buttermilk.[1] In hotter climates moss might dry off during summers, and comes back to "life" during seasons with precipitation.

Wild spaces in cities are juxtapositions of natural and built heritage; which in the case of Kolkata and Bengaluru have represented eras foregone. Sometimes eras so old, that even logical constructs of the purposes

DOI: 10.4324/9781003374602-2

of the human-made structures have become difficult to draw without sufficient architectural, historical, and demographic knowledge. Within this gap in understanding between "what was" and "what could have been", lie potential constructs of landscape fractions. It is important to understand these compositions – not just for their values of art and aesthetics, but also as markers of time and age.

The plant-built compositions have the capacity to capture, and often create, an impression of time – an essence that will no longer hold, if the same plants or human-made structures are separated from each other. Plant growth offers a notion of "old age" in character. David Watt, a chartered surveyor working with historic buildings based outside of the United Kingdom, refers to this enhancement of character as "historic and aesthetic value" (Watt, 2006). One can relate such values of appeal and memory to Austrian art historian Alois Reigl's framework of critiquing architecture which I explore in more detail, later in Chapter 4 on Aesthetics. While a sense of age from moss and lichens growing on buildings is charming, it is also misleading as it raises the question of "how old?" to an observer. Reigl suggests the term "pastness" since a sense of age can be fabricated using clever manipulation of objects' appearances and their discourses (Holtorf, 2017). Indeed then, botanical growth over human-made relics cannot always be held as markers of age, but rather to evoke a perception of it. This is a value I also discuss more in Chapter 4 on Aesthetics.

Found in the urban-nature assemblages are not just buildings that are from a foregone era, but also relics that are modern. Technological equipment or machines abandoned within a factory yard, or broken automobiles along the roadside covered with plants are also relics whose logical counterparts as Gandy (2005) mentions belong elsewhere: abandoned cars that belong on the road, in parking lots, in showrooms for sale, or even private garages; or abandoned manufacturing machines that would otherwise form a critical part in the production line along with other machines in the factories. Perhaps it makes little sense to rate these objects for their original purpose, when they lie displaced from their usual context. This illogic is not just applied to human-made objects, but also plants that are not expected to grow on buildings. While a sense of age can be derived from plants, the era is defined together by relics, that are now intertwined with the plants. These ensembles are clearly composites that are capable of inciting new meanings when viewed as a whole. Otherwise, if the plants were to be separated from machines, the composite would be broken down to industrial legacy on one hand, and nature on the other. More intuitively, when separated, the two elements do not appeal to age but rather technology and its absence.

The relics could be architectural elements built 50 years ago or even e-wastes from only five years ago. Residual spaces along roadsides can become dumping grounds for the abandoned automobiles. Overgrown grass and vines represent the long period since when the owners of these machines had discarded them. I found one such assemblage along a highway in Bengaluru: a recent model of a motorbike that lay abandoned along the roadside with grass growing around it. Scraped paint and rusted body parts of the machinery made it look older than it actually was. The assemblage of a colour-scraped motorbike with grass, as a whole, makes a vintage appeal depicting a false notion of old age; but will at least serve as a marker of industrial legacy. Elsewhere, I found, older models of motor vehicles which looked new as their body paint was intact. This is how the sense of pastness gets fabricated.

One can not only see spontaneous nature growing on buildings in the traditional areas of a city, whose fabric dates centuries old, but also in neighbourhoods that are built as recently as in the last few decades. While the fractions of historical architecture in Kolkata can demonstrate a mix of elements from Indian, Mughal, and European architecture – the more recent assemblages in cities are comprised of steel frames and concrete aesthetics. Take for instance the locked-up bicycle stand in Bengaluru, where wild plants have grown without human care – its visual character shaped by the meshed metal frame, postmodern buildings, and stained soil from oil spills from vehicles that use the ground in front for parking (Figure 2.1). This represents technology of a more modern age; carrying elements from the result of recent industrial revolution into the post-industrial times of today.

Indeed, it is not uncommon for landscape fractions in cities to carry remains of influential architectural legacies e.g., from post-industrial and colonial eras. However, the disappearance of such assemblages with high heritage value is also common. People may decide to tear down old industrial structures which have fallen out of use and instead rebuild the space for more contemporary purposes. But other times, monumental buildings can be retained as a base to build new experiences upon. A movement has begun to preserve and reuse sites of industrial accomplishments, by turning them into public and cultural spaces (I discuss some examples in Chapter 4 on Aesthetics). Notably, the interesting

FIGURE 2.1 Visual character shaped by architectural legacies: wild growth inside the locked-up bicycle stand in Bengaluru. July 26, 2015.

settings of wild growth in these sites that have come into existence by chance, is not always preserved, and landscapes around the industrial monuments are rather redesigned – dismantling the aesthetics and experiences of built-nature assemblages that have formed spontaneously.

Cyborg assemblage

In the era of Anthropocene, as the world moves towards blurry lines of what is "built", and what is "natural", cities show traces of how there can be assemblages that are both built and natural. These assemblages are fusions of living and non-living, and can be viewed as cyborgs. Donna Haraway's essay, *A Cyborg Manifesto* describes a cyborg as "a cybernetic organism, a hybrid of machine and organism, a creature of social reality as well as a creature of fiction" (Haraway, 1991). The author argues that, with the proliferation of technology in our lives, humans have become chimeras (or hybrids of machine and organism). The earliest memory of mine that vividly captured this image of a cyborg is perhaps the fictional[2] character of RoboCop: a human cybernetically enhanced with bionic parts. In reality, prosthetic and bionic limbs are not uncommon, and are examples of less extreme transformations or even extension of the human in comparison to the sci-fi character. Similar transformations have also been successfully made on non-human life e.g., technology is retrofitted on to wild animals and pets (e.g., for tracking or prosthetics) or on to plants (e.g., to act as sensors).[3] Haraway's conceptualisation of a cyborg dilutes several hierarchies including that of the animate and inanimate as well as nature and machines; reflecting very much the characteristics of fused assemblages of plants and buildings that challenge the dominant hierarchy of the built over natural within cities. Among architects, Le Corbusier's reference to a house as a machine has been known since the 1920s (Corbusier, 1923). Conveniently, I extend this imaginary of buildings as machines and substitute the human from RoboCop with plants, to construct meanings of more-than-human portrayals of landscape fractions; claiming them as cyborgs.

The city is riddled with such chimeras, such cyborgs, such unplanned yet dramatic botanical growths which are seemingly inseparable from the built surfaces – plants clenching on to the buildings, in aesthetic displays of art composed by forces of nature and human beings. Perhaps it is odd to picture how plants – lifeforms incapable of voluntary motion in a full sense – could re-wire themselves so easily around the buildings (and human altered spaces) while most other non-human life (animals) struggle to do so. In such mergers of life and technology, the plant-built assemblages are where one can find the trees bury its roots deep inside the buildings. The trees that grow out of the pores of roofs and walls, using the decaying masonries as a growth medium, in that sense, become extensions of the building. These cyborgs are more than just material arrangements bound by biological processes and are socially produced through discourses that frame and construct symbolic meanings of these assemblages (see Chapter 3 on Existence and Chapter 7 on Frames). Erik Swyngedouw in his essay *Circulations and Metabolisms: (Hybrid) Natures and (Cyborg) Cities* notes such intermingling as "socio-environmental milieu that welds nature, society and the city together" (Swyngedouw, 2006, p. 105). Swyngedouw also points to how this happens through multiple layers of "networked technostructures" that may include piping, ducts, communication infrastructures, transportation apparatuses, and so on (p. 105). In a systems view, such layers are also understood as "infrastructures of human and non-human systems" and "[infrastructures of] living and non-living entities" (Lokman, 2017). Such infrastructures may be visible overground or hidden underground facilitating growth of spontaneous nature in cities.

Spontaneous nature may also be referred to as (informal) green infrastructure, but which are disconnected from processes that integrate parts of the city to perform as a unified system. Fractions of botanical growth growing on top of built surfaces are isolated islands where plant communities draw their uniqueness from a host – the human-made objects – through the process of symbiosis. They are not always ordinary arrangements of plants that draw their nutrients from the soil, but rather from a range of other intensely modified growth-mediums of the built environment. Broadly speaking, biological growths of any kind are typically viewed as part of the greening movement and as our armament against climate change; this notwithstanding, not all natural growths are desirable. So, one is tempted to relate landscape fractions to green infrastructure. However, the fundamental problem that limits the construct of this meaning may be an assemblage's apparent isolation from the city's metabolism and circulation.[4] Unlike designed landscapes and buildings which are finally inter-connected by a network of physical infrastructures like roads, pipes, and cables – fusions of plants and buildings seem to be largely, if not completely, disconnected from the machinery (or networked technostructures) of the city. For instance, while parks and roadside landscaping may be connected to sprinkler systems that use piped water or are fed through tanker trucks carrying water, an assemblage of strangler plants growing over a building draws its nutrients from the masonry itself. Or say, for instance, the municipality would have disconnected the piped water supply to abandoned plots or supply of fertilisers to parks. Cut off from the regular flows of materials and nutrients that are otherwise intended to support plant-life, the sites of cyborg assemblages are merely local, at least in terms of the resources they depend on. Therefore, they can be pictured as severed from urban metabolism. This notwithstanding, there are a few exceptions like incidental conditions for plant growth facilitated by water made available from leaking underground drains or water supply pipelines; or nutrient rich soils flushed away from elsewhere due to flooding.

Isolated from the city's integrative processes, landscape fractions of spontaneous nature are not recognised as public assets. This disconnect of cyborg assemblages from the network of mechanical flows as well as their disregard in policy making and built-environment design can perhaps be linked to Foucault's analogy of antipathy. Antipathy according to Foucault (2005) "maintains the isolation of things" (p. 27). Foucault explains antipathy as animosity or hatred among beings – a prey and its predator for instance – that prevents assimilation and preserves diversity to allow different things to continue existing in their authentic and unique nature. Despite this understanding of antipathy as nature's way of allowing differences to exist, one problematic predicament is that isolation of people and nature, manifested as emotional disconnect between the two, can become responsible for the lack of vigour in pursuing environmental goals even though they are essential for a society to sustain (Whitburn, Linklater, & Milfont, 2018; Kollmuss & Agyeman, 2002). Then antipathy is not entirely a desirable trait when considering the citizenship of wild spaces; but perhaps sympathy for these ensembles may trigger the opposite[5] – encouraging pro-environmental attitudes among urban residents.

Life and death

To begin, I reiterate two salient points mentioned before in this book: a) landscape fractions composed of spontaneous nature draw their meanings from people's perceptions of wild growth; b) people are found to be ambivalent of such landscapes. Spontaneous nature faces some criticism of dereliction and decay, but are largely ignored and eventually disposed of when structures and spaces go through the process of renewal. This is because, fractions made up of spontaneous nature, are thought to be temporarily existing in cities – meaning they are there, momentarily amid the ruins, until the site is re-built, or the property is renovated. The construct of their meanings as temporary or transient is perhaps one reason why very little work has been done to understand these natures. Respecting the fact that more research needs to be done, my goal is neither to criticise nor to dispose of the assemblages made from wild plants and built-environment; and instead to explore their ambivalence a bit more.

Ensembles of spontaneous growth can be viewed in terms of life and death, organisms and machines, or transient and stationary. On one hand, several scholars have discussed meanings of spontaneous growth on buildings as signs of urban decay (Del Tredici, 2010); and urban decay can also be understood as a city moving towards its death – in whole or in parts. Contrary to this perspective, biological growth on buildings has also been noted to add age value to buildings – which add more years to the perceived lifespan of the building. Take for instance moss growing on buildings which, as Watt (2006) has pointed out, adds to the historic value of buildings. So, on one hand, vegetation growing on built surfaces can signify "the end", and on the other, it can also signify the timelessness of the buildings. Death and timelessness are two very contrasting and opposite ends of the spectrum that forms the continuum for people's perception of landscape fractions. Discussing the "living and dying" of a city, a paragraph from Ash Amin and Nigel Thrift's book Seeing Like a City discusses the fate of people in the hands of city components such as tramlines. Amin and Thrift (2017) write:

> *because each of the tramlines of infrastructure contains its own peculiar forms of cruelty as well as promise. We use the word cruelty knowingly, since we are talking here about machines that legislate who and what lives and who and what dies, and who and what lives in what form.*
>
> *(p. 23)*

But unlike death which can come from an uprooted or failed tree, strangler plants have been argued to, but only at times, strengthen otherwise failing structures (e.g., wall trees of Hong Kong mentioned later in Chapter 3 on Existence making dilapidating structures more immune to structural failure, whereas smaller plants like moss and ivy are naturally known to pose no particular threat to the built structure. So, the symbiosis of the "cyborg structure", is beneficial for the sustenance of both the organism and the building. If the landscape fractions are comprised of plants over human-made infrastructure, then at least the plants bring life; which is contrary to Amin and Thrift's account of death by infrastructure.

Given the intense artificial processes that go into creating urban nature, city dwellers often do not get to witness the full biological cycles of nature. Trees and flowers are grown in nurseries, and planted in public gardens after they have crossed the initial stages of growth. On one hand, while human control can make landscaping become more resilient, it also veils the delicate processes of seeds growing to become saplings, and then into mature trees. On the other hand, this is a distortion of natural processes itself – treescapes do not appear in nature overnight; they typically follow different stages of ecological succession. These stages of succession are better witnessed in abandoned plots, unmanaged backyards, or even small patches of growth along the roadside and on the derelict buildings, where first there emerges ground cover, then there appear shrubs, and finally there are taller trees. Spontaneous settings – even in small quantities – can become spectacular sites that demonstrate how nature evolves over time; and how dirt transforms into new life. This temporal experience is grounded in real-time transformation of landscapes and serves as educational material beyond textbooks and audio-visual documentaries that are either snapshots or fast forwarded using time-lapse techniques; the latter involving photographic documentation provides evidence, but not without disorienting the observer's sense of time and rhythms that create nature.

At the same time, sites of urban decay composed of human-made relics in nature or natural growth on the relics have been for a while, of interest to photographers and filmmakers. They capture these sites while referring to them as "abandoned landscapes" and "decay photography"; almost branding the degenerative aspects of abandonment and decay as art. One photographer, Dennis Thomas, considers photographing urban decay as a challenge to make the sites of decay appear aesthetic through patterns, textures, and colours (Thomas. 2014). Thomas expresses decay photography as "ability of find the beauty in the blight", and notes how the challenge of "creating captivating compositions from scenes of collapsing architecture" have particularly drawn interest of budding photographers (Thomas, 2014, p. 2). This is not surprising given sites of dilapidating architectural settings or other relics that symbolises decay are both abundant and yet mysterious to pique the interests of urban explorers.

The process of interpreting wild spaces appears to be intuitive and enshrined in values that relate to rarity in thought. I have found it hard sometimes to decide if compositions of spontaneous nature are worth my interest, yet such arrangements are relatively easy to find when compared to some other subjects of photography e.g., migratory birds appearing for a few days in the city to weather events that last for a few hours. Despite this abundance, human-nature relics are "strange things" that may act as puzzles left to the observers to determine how these relics have come into existence. Relics do not necessarily have to be buildings depicting an era in time. They could as well be technological relics as small as a bicycle to as large as a road truck – that are not abandoned,

FIGURE 2.2 Signifying both life and death in the same scene, forces of Gaia can be felt in trees reclaiming ghosted buildings along Ramdulal Sircar Street in Kolkata. July 14, 2018.

but just parked amid wild spaces. Then, I can perhaps say, landscape fractions emerge out of an interplay between intuition and logic created by (temporary) aesthetic arrangements of human-made objects and nature.

Spontaneous vegetation growing on built surfaces can signify new life. Natural ecosystems altered heavily by urbanisation cannot revert back to their original forms, and instead nature has to adapt to built-environments to re-establish its character. Landscape fractions can also be viewed from the position of James Lovelock's hypothesis, for trees growing on buildings can be visualised as forces of Gaia reclaiming their right to the land which was taken away by near-impervious built surfaces (Lovelock, 2000). However, it is not at all uncommon to view the same landscapes as outcomes of human neglect rather than as force of nature. A partially abandoned building on Ramdulal Sircar street in Kolkata is now stripped of its wall plaster exposing the red-brick masonry (Figure 2.2); crevices of which now become a catching bed for bird droppings and seeds carried by the wind. The mud-baked roofing tiles that once shielded the rooms for inhabitants, has now disappeared with time, to lay bare its wooden truss that formed the frame – the decay of roofing technology allows light, rain, and air to create conditions for biological growth to flourish. Not surprisingly, the otherwise dilapidating structure can be seen to have become habitat for banyans and other plants.

Scarcity of resources

Spontaneous vegetation becomes a mode of landscaping in communities lacking resources such as land, water, and management staff. In cities, non-human nature would seem to be in competition with human society when large volumes of water are used for maintaining parks and gardens in an area, while somewhere else, there may be households without adequate water supply. In deprived areas, the city might have had to prioritise household water by compromising water supplied to maintain parks – as a result, the plants in the parks might be drying and ground cover might have all died out to expose the bare soil. At the same time, due to austerity cuts, such areas may also be under neglect from maintenance of their public infrastructure – cracks on the road, leaking joints of underground pipes, or weathering plaster of storm water drains are enough to create conditions for biological growth. This book is replete with examples of such nature that grows over pedestrian walkways, inside drains, and over leaking sewerage joints, often hidden from everyday view.

The hand-drawn rickshaw parked in the street in one of Kolkata's former suburbs – now engulfed by urban expansion – belongs to a resident who ferries people for short distances and receives a nominal fare. Like most organic extensions in a city, the rickshaw-puller's neighbourhood has also emerged with narrow (unplanned) streets. Such streets are generally considered space-constrained making it difficult for planning even basic services such as access to properties, water and sanitation network, and so on. Nature is not a priority here, and conventional green infrastructure such as an avenue of large trees certainly needs more space to grow than is available. Some landowners who can afford space have planted tall coconut trees. But not all arrangements of plants consume more space and, residents have also planted small trees, shrubs, and creepers, which give the street otherwise devoid of greening initiatives, a visual relief.

In the traditional part of Kolkata, often referred to as "Old Calcutta", the built fabric is dense, with narrow streets. Having less gaps between buildings helps keep the built-environment cool during summer – as it reduces exposure to the sun. The tall buildings offer shade and cool pathways for thoroughfare amid an otherwise hot climate. But this also means dark alleys with virtually no space for nature. It means, while there is ambient temperature, there is no "ambient nature" that can serve as visual relief. In one such alley, a bed of moss has grown from decay of a built-surface (Figure 2.3). The seeping rainwater coupled with a low exposure to sun has created moist conditions for this growth. There are other plants that have also emerged; ferns are evident. The assemblage is tucked inside a vast and dense network of narrow streets, and notably not all the streets have such nature growing in them.

Landscape fractions are small scale nature and would add to environmental justice if presence of such assemblages can bring benefits of nature to people who otherwise do not have access to nature. This

FIGURE 2.3 Spontaneous nature as mode of greening: moss and lichens add visual relief along a narrow alley with a residential neighbourhood of Kolkata. July 14, 2018.

approach of allowing minimal nature in neighbourhoods that are otherwise deprived of more flamboyant green spaces, follows what Jennifer Wolch, Jason Byrne, and Joshua Newell advocate as "just green enough" (Wolch, Byrne, & Newell, 2014). Dense neighbourhoods often do not have roadside trees, let alone parks. The prime resource in Kolkata and Bengaluru is space, as vacant plots or even residual spaces are typically scarce in the densely built traditional areas; as opposed to more recently planned areas. In high-density cities, allocating adequate water for public landscaping can also become a challenge because of high water demand for other purposes. However, landscape fractions seem to have been created through plants defying these resource concerns and adapting to the urban conditions easily, without needing additional resources. Assemblages that obey economy of space, such as plants growing out of service ventilators or on rooftops, are a common sight. Furthermore, creepers and moss can both offer a green film over surfaces creating a refreshing green ambience – it is not equivalent to compensate, but at least a consolation, for the absence of plants in the alleys.

Spontaneous landscapes in cities are also seen as reflections of the city's inability to maintain designed landscaping. This is due to the failure of particularly capitalist regimes of greening which carries the assumption that harbouring nature in cities requires resources to be expended by humans, which can happen at the cost of other aspects of human development. For instance, some neighbourhoods are faced with low allocation of public funds to maintain parks and gardens which provide access to nature. The funds instead are prioritised to ensure further development of economically better-performing districts within the city. To reiterate from Chapter 1 on Morphology, Del Tredici (2010) highlights how in depressed cities spontaneous nature have attained stable multi-layered forms in spaces right within the city cores which remained abandoned and neglected for a long time. I will return to this point when explaining the formation of these ensembles in Chapter 3 on Existence. But here the question I am trying to pose is: what then do these wild spaces mean for communities for whom arrangements of landscape fractions are part of everyday nature? Bradshaw, Linneker, and Lundy (2020) point to how economically disadvantaged communities become affected by low access to the typical natural capital enjoyed by other communities in a city.

Small scale spontaneous nature is often linked with marginal spaces or even decay from lack of care and upkeep. In such cases, minimal nature can be a response where spaces have been decoupled from the economic growth of the city. It is worth noting, that in the past, some processes like "guerrilla gardening" have created nature (taking full advantage of marginal spaces) by forces of socialism. The *Organopónicos system* in Cuba unlocked rooftops, vacant buildings and other marginal spaces which were used for growing food (Ewing, 2008). Here, small-scale green spaces were produced deliberately, not for visual aesthetics but for utilitarian purposes. Many years after the crisis in Cuba, one can continue to find similar movements without any link to a war emergency (Hardman, et al., 2018).

A row of houses has been built somewhere in Kolkata, within a 30-minute travel towards north of the old city area. One can notice that although the roads are well laid out and wide, houses here are modest with tiled roofs as opposed to reinforced concrete. The spill over of everyday activities outside the house reflects a certain lack of space

FIGURE 2.4 Pumpkin vines intentionally grown on a rooftop in Kolkata, show how small-scale assemblages with little or no maintenance can cater to food supply within cities. April 14, 2018.

indoors. Together, these aspects reflect that this residential colony has been planned for offering affordable housing; economising the utility of spaces and built-structures. More affluent neighbourhoods would be envisaged with green spaces, but roadsides here are cemented to avoid maintenance costs. However, this reflects more of the architect's conception of the space and not people's preferences; as the Indian parakeet, a popular pet-bird, can be seen in a cage of a typical house (Figure 2.4) that signals a strong affinity of the dwellers towards nature. There are also potted plants that add to greening of the locality.[6] The typical house has pumpkin plants growing on the rooftop. Pumpkins are edible not only as full-grown vegetables, but also pumpkin flowers may be fried with gram flour for consuming as a side dish. The climber plant not only adds green cover in the area, but also serves as means of food supply. It helps one think how a variety of vegetation – grasses, moss, vines, trees – can grow spontaneously imitating designed settings on the rooftops; and the significance of such vegetation may range from sources of food to functions of cooling. The example of growing pumpkins on rooftop is also one of subsistence i.e., due to its small portion, households typically use such produce for their own consumption. Later in Chapter 5 on Collectives the book illustrates through grazing of cattle, how people also use landscape fractions to enter the mainstream capitalist economy.

Given such benefits within an overarching regime of failure, one can be faced with multiple dilemmas of how to discern utilities of such small-scale landscapes for urban planning and greening policy – i.e., do we retain them considering benefits they bring to communities? Or do we elevate conditions of decay of which landscape fractions are only symptoms? Although small-scale wild urban spaces are part of everyday nature and play provisioning (e.g., food provision) and cultural roles (e.g., visual relief), are they green enough to play the regulatory and supporting roles (such as improving air quality or soil formation)? And so on.

Threat to human and non-human life

The stranglers and grasses growing on buildings had typically emerged from the cracks. The cracks could have resulted from a natural withering of building, from root action pushing through the cervices of masonry or from the sheer weight of the giant stranglers that grow in unexpected positions on top of the built structure. Sometimes it is a combination of all of these compounded by harsh weather such as rain, heat, or cold. The inflicted damage to buildings may result in loss of structural integrity and are serious hazards. Without management, structures have been documented to collapse, and hurt people who might have taken shelter under them or who were walking by roads abutting structures. Iron rods sticking out because of weathered reinforced concrete, rusted from rain and moisture can become dangerous for those using the site. Glass pieces, nails cloaked by plant growth can also be injurious to unaware trespassers – both human and non-human.

Wild growth bearing interesting life, can become inviting to visitors, while concealing certain dangers. Bright red flowers had grown in the small front yard of an abandoned house – a house that had its window panes missing from the frames, and an empty cavity on the wall, where the door was supposed to be fitted. Clearly, the building was abandoned midway during construction. Like the neighbouring houses that were all completed projects, perhaps the abandoned house too had another floor planned. Yet the roof was left unfinished with no parapet and no drainage channels. The rainwater pooled on top of the roof creating perfect conditions for grass and other small flowering plant species to emerge. Together the water and plants made a habitat for drawing dragonflies and other insects. The easy access to the property and unsupervised fun, invites both children and adults who use the space at different times of the day. But at the risk of interacting with the dilapidating conditions of the building and possible encounters with snakes or feral animals that may be taking shelter within. Similar settings along the road of, say, exposed electric cables cloaked by ivy, or abandoned equipment rusting beneath the wild growth, can pose no small risk to human health.

Small-scale spontaneous nature in cities exists in close proximity to human beings. They grow under the overhead water storage tanks, whose structure may have become weak due to aging concrete, giving way to leakages. They grow along busy roadsides and railway tracks where accidents with cars and trains are common. They grow on abandoned houses and dilapidating public buildings which are no longer fit to occupy and have therefore been decommissioned from use. They may grow upon retention walls abutting property alongside a busy road; whose foundations may have grown weaker due to waterlogging in soil and the growing weight of plant growth; posing risk to pedestrians. In other instances, small-scale nature may be composed of plant species that are able to cause discomfort upon contact.

The dilapidating structures put at risk, hidden lives of urban spaces around assemblages of nature. Abandoned houses are easy target that illegally serve as temporary shelter for homeless. Occasional visitors are also those who carry out precarious activities such as gambling, drinking, drugs, and prostitution. Other times, these spaces may even become infrastructure for inquisitive adults triggering creative imaginations such as in ghost walks or become children's playgrounds for the game of hide and seek.[7] Dilapidating and abandoned structures are also animal habitats; encouraging owls, rodents, bats, snakes, mongooses, and so on (see Deb, 2017). In some cases, dilapidating structures are marked with bold signage cautioning people to stay away. However, in my fieldwork, abandoned buildings I found were often left unmarked (and unmonitored); perhaps with the assumption that they offer no reason for people to trespass.

On the other hand, unlike nature growing on buildings, wild growth growing on the ground – say along the roadsides – cannot be associated with structural failure that can directly affect human life. In that sense, they can be considered relatively less harmful. Small patches of wild growth typically do not pose health risks to people in its surroundings. This notwithstanding certain plants can cause allergic reactions and skin irritations.[8] As one example, the Indian sting nettle (*Tragia involucrate*), is a herb that is found in several parts of the country. An otherwise useful herb, with health benefits when parts of this plant is processed for consumption, is also well known to cause skin rashes and irritations upon touching with bare hands. Such wild plants may also have harmful effects for animals. Take for instance the case of parthenium, an invasive species of plants rapidly colonising the urban landscape in India. Ignoring the obvious disadvantages of invasive proliferation, the tiny white flowers of the plant to me, at first, appeared quite appealing – as if the daylight reflection of a starry night embedded on the ground. They are so ubiquitous that it is rather common for cattle to come in contact with them when they stray around neighbourhood streets and open wild spaces for grazing. Yet, parthenium does not serve as fodder for cattle, and research has also suggested that interactions of cattle with the plant may be responsible for causing udder inflammation, rashes, and fever (TNAU, 2016). Later, in Chapter 5 on Collectives, I discuss how spontaneous patches also serve as grazing grounds; this notwithstanding certain plants – such as parthenium and stinging nettle – are not generally consumed by cattle.[9]

Human experience with nature may shape people's affinity towards spontaneity in nature. The rapid spread of parthenium is considered problematic worldwide (Patel, 2011). Sometimes, spontaneous landscapes become only dominated by a few such species; which can reduce the meanings associated with these landscape fractions. Wild growth in a locality can become synonymous with one or two plants that are harmful, and people may tend to develop a negative perception of all wild growth. However, there are other plants that are harmless and even add to a place's aesthetics more. A stigma around spontaneous nature can narrow people's acceptance of wild growth, and in turn citizens may end up adopting heuristics such as paving and weed control for all wild growth on mass scale. Chapter 7 on Frames will delve more on how meanings of landscape fractions get produced through context in which they are placed.

Notes

1 H. Vangara. Effects of plant growth on buildings, interview with A. Deb, January 11, 2022.
2 RoboCop is an American sci-fi movie released in 1987.
3 Experiments are being conducted on plants to induce mechanical movements. Read more about the project "Cyborg Botany: Augmented plants as sensors, displays, and actuators" at https://www.media.mit.edu/projects/cyborg-botany/overview/
4 Building on works of Marxian scholars (including Marx himself), Swyngedouw (2006) takes "metabolism" and "circulation" as entry points to analyse "socio-natural things". He describes metabolism as, exchanges of matter between organisms and the environment that includes bio-physical processes of both living and non-living entities; and circulation as flow of commodities and capital.
5 Foucault (2005) considers sympathy and antipathy, as twins that compensates for each other.
6 Plants and birds form part of human well-being; and perhaps hinting towards the Biophilia Hypothesis of naturalist E.O. Wilson that suggests humans are hard-wired in their brains to be drawn towards nature. A more mathematic approach has been taken by environmental psychologists Rachel Kaplan and Stephen Kaplan to confirm health benefits of people's proximity to nature (see Chapter 5 on Collectives).
7 Katy Mugford in her book chapter within Anna Jorgensen and Richard Keenan's edited book *Urban Wildscapes*, discusses the benefits of unsupervised play such as 'I-spy' for healthy child development (Mugford, 2012, p. 83).
8 J. Munirathinam. Ecology of small-scale wild growths in cities, interview with A. Deb, November 23, 2021.

9 J. Munirathinam. Ecology of small-scale wild growths in cities, interview with A. Deb, November 23, 2021.

References

Amin, A., & Thrift, N. (2017). *Seeing Like a City*. Cambridge: Polity.

Bradshaw, S., Linneker, B., & Lundy, L. (2020). Naturally feeling good? Exploring understandings of 'green' urban spaces in the Global South. In N. Dempsey, & J. Dobson (Eds.), *Naturally Challenged: Contested Perceptions and Practices in Urban Green Spaces* (pp. 37–57). Switzerland: Springer.

Corbusier, L. (1923). *Vers une Architecture (Toward an Architecture)*. Paris: Les éditions G. Crès.

Deb, A. (2017). Understanding urban 'wildlife'. In A. Deb, *Introducing Wildlife in Urban Ecosystems* (pp. 13–43). New Delhi: Copal Publishing Group.

Del Tredici, P. (2010). Spontaneous urban vegetation: reflections of change in a globalized world. *Nature and Culture*, 5(3), 299–315.

Ewing, E. (2008, April 4). Cuba's organic revolution. Retrieved January 5, 2023, from *The Guardian*: https://www.theguardian.com/environment/2008/apr/04/organics.food.

Foucault, M. (2005). *Order of Things* (Reprint ed.). New York: Routledge.

Gandy, M. (2005). Cyborg urbanization: complexity and monstrosity in the contemporary city. *International Journal of Urban and Regional Research*, 29 (1), 26–49. Retrieved from http://www.matthewgandy.co.uk/datalive/downloadfiles/ijurr.pdf.

Haraway, D. (1991). A cyborg manifesto: science, technology, and socialist-feminism in the late twentieth century. *Simians, Cyborgs and Women: The Reinvention of Nature*, pp. 149–181. Retrieved June 1, 2022, from http://www.faculty.umb.edu/gary_zabel/Courses/Art%20and%20Philosophy%20in%20SL%20and%20Other%20Virtual%20Worlds/Texts/cyborg_manifesto.pdf.

Hardman, M., Chipungu, L., Magidimisha, H., Larkham, P. J., Scott, A. J., & Armitage, R. P. (2018). Guerrilla gardening and green activism: rethinking the informal urban growing movement. *Landscape and Urban Planning*, 170, 6–14. doi:https://doi.org/10.1016/j.landurbplan.2017.08.015.

Holtorf, C. (2017). Perceiving the past: from age value to pastness. *International Journal of Cultural Property*, 24(4), 497–515. doi:10.1017/S0940739117000236.

Kollmuss, A., & Agyeman, J. (2002). Mind the gap: why do people act environmentally and what are the barriers to pro-environmental behavior? *Environmental Education Research*, 8(3), 239–260. doi:10.1080/13504620220145401.

Lokman, K. (2017). Cyborg landscapes: choreographing resilient interactions between infrastructure, ecology, and society. *Journal of Landscape Architecture*, 12(1), 60–73.

Lovelock, J. (2000). *Gaia: A New Look at Life on Earth*. New York: Oxford University Press.

Mugford, K. (2012). Nature, nurture; danger, adventure; junkyard, paradise: the role of wildscapes in children's literature. In A. Jorgensen, & R. Keenan (Eds.), *Urban Wildscapes* (pp. 80–96). New York: Routledge.
Patel, S. (2011). Harmful and beneficial aspects of Parthenium hysterophorus: an update. *3 Biotech*, 1(1), 1–9. Retrieved from https://doi.org/10.1007%2Fs13205-011-0007-7.
Swyngedouw, E. (2006). Circulations and metabolisms: (hybrid) natures and (cyborg) cities. *Science as Culture*, 15(2), 105–121.
Thomas, J. D. (2014). *Urban and Rural Decay Photography: How to Capture Beauty in the Blight*. Burlington, MA: Focal Press.
TNAU. (2016). *Parthenium on livestock health*. Retrieved April 13, 2022, from TNAU Agritech Portal: https://agritech.tnau.ac.in/agriculture/agri_weedmgt_parthenium_livestock.html.
Watt, D. (2006). Managing biological growth on buildings. *Historic Churches. The Building Conservation Directory: Special Report Magazine*, 13, pp. 36–38. Retrieved from https://www.buildingconservation.com/articles/bio/bio.htm.
Whitburn, J., Linklater, W. L., & Milfont, T. L. (2018). Exposure to urban nature and tree planting are related to pro-environmental behavior via connection to nature, the use of nature for psychological restoration, and environmental attitudes. *Environment and Behavior*, 57(7), 1–24.
Wolch, J. R., Byrne, J., & Newell, J. P. (2014). Urban green space, public health, and environmental justice: the challenge of making cities 'just green enough'. *Landscape and Urban Planning*, 125, 234–244.

3

EXISTENCE

Outside human control

To me at first, it was a mystery as to why some small-scale nature that has come about by chance, continues to survive whereas others do not. I questioned why fig trees for instance, that had begun growing on buildings, remained untouched by property owners? Other times, it was hard to fathom why an entirely unpaved and moist street-side would only have grasses growing in one small patch – rather than along its entire stretch? Sometimes I wondered how grass-beds had managed to emerge over a rooftop, without soil matter or a steady supply of nutrients? Thus, I find two streams of enquiry as essential: the way landscape fractions are created (or how they come into existence); and the way they survive (or how they continue to exist). The spontaneous arrangements of nature show how cities have given rise to distinct assemblages of social, physical, and nature. Uniqueness of wild spaces within cities are a reminder of the fact that even when left to forces of natural regeneration, urban ecosystems have reached a point of no return, and may be referred to as novel ecosystems. See Chapter 6 for more.

Inside cities, where nature's autonomy is weak, human decisions can lead to the rapid transformation of the urban forests, and at a very large scale. For instance, some local governments have demolished buildings within cities to replace them with green and blue infrastructure.[1] Similarly,

DOI: 10.4324/9781003374602-3

spontaneity of nature in cities, too, can be shaped by various instruments of public policy and planning. On one hand, cities may declare "weed ordinances" that can encourage more engagement with the vegetation to control their growth (see Gandy, 2018). On the other, planning authorities may deliberately avoid management to allow naturalistic growth; for instance, Sheffield city authorities have announced that they will avoid trimming roadside verges to allow for more biodiversity.[2] The important thing, perhaps, to note is that such high degree of control gives urban nature a reputation of being "transient" (see Deb, 2017). Such rapid and rather instantaneous changes to construct and destroy landscapes, thwart traditional processes of natural evolution within cities; but despite this control, unplanned nature continues to find ways to survive in urban spaces. Wild plant-growths in cities have been able to establish alliances with the existing layers of the built-environments as well as flows of resources and nutrients within them – even though these human-designed systems were originally put in place to move away from the natural. The dysfunctionality of systems – such as overflowing water from buildings on to the streets, exposed masonry of dilapidated buildings, leaking underground water pipes, and so on – create conditions for growth of spontaneous nature.

To build a more meticulous picture, imagine this scenario: the local government of a city decided to block a road for the purpose of repair. Behind the cautionary "road closed" or "men at work" signs, a common sight are the gigantic cranes that excavate roads as well as roadside verges to lay pipes, cables, and other sub-surface infrastructure. The hard and compacted urban soil can be rather difficult to penetrate through more rudimentary tools, yet an army of engineers and on-site workers fill the street to monitor the construction work. They all ignore the street-side nature, because nature can be regrown here, once the civil works are done. Moreover, the scale of the task is huge and the excavation works run for kilometres; there is simply no time to pay attention to such details. As a result of the excavation, not just tar and paving, but all vegetation that get in the way, would also be removed for now. Ultimately, the excavated soil will be put back again to cover the deep trenches. Revegetation can be possible almost immediately: this can happen through planting saplings and monitoring them as they grow; or planting more mature species to turn the street verges into a pleasant landscape overnight. But one predicament that blurs this potential is that, if project resources do not permit, revegetation will not be part of this operation. Consequently, there may be a delay in planned landscaping; allowing sites to revegetate themselves naturally.

Spaces of spontaneous nature challenge the widely regarded notion of urban areas as "human-dominated systems".[3] Such a designation speaks of the high degree of control and creativity which allow humans to craft nature into desired settings in urban surroundings. Alternatively, in some cases, wild landscapes can be found to compete with the aesthetics of designed landscapes. Noting how beautiful landscapes in cities also exist in forms unintended by humans, Matthew Gandy describes experience of such *unintentional landscapes* as "an aesthetic encounter with nature that has not been purposively created" (Gandy, 2016, p. 434). Gandy explains that the process of drawing aesthetics relies upon a sort of enframement in the human mind that is not necessarily guided by visual appreciation alone. He finds such landscapes in neglected spaces which have been produced alongside anthropogenic activities in the era of globalisation – thus framing the explanation with sensory experience, production of space, and a time-period in human civilisation.

When humans didn't intend it, there must certainly have been agencies – outside of human intention – responsible for its creation. To clarify, agencies outside of human intentions, can still be humans, but those whose conscious effort did not, or could not, affect the unintentional natural settings. For instance, an owners' decision to stay away from his house for a long period could have helped develop the tree roots that had penetrated his building walls and roofs to create an interesting assemblage – notwithstanding such biological growth on buildings are at times incompatible with the intended purpose of architectural designs (see Figure 3.1). Meanwhile, another house owner with plants growing out of his façade would have found it difficult to pay the maintenance costs after years of neglect that was caused by, perhaps, the tenants who inhabited the space. Migration and economy are only two of the presumably many forces which act to create interesting novel landscapes – not purposively created by humans. Thus, before declaring nature in cities as human-dominated, it becomes important, and even fascinating, to recognise that factors responsible for the existence of unintentional urban nature are in fact outside of human control.

There needs to be more clarity in use of the phrase "outside human control" for to distinguish the natural and spontaneous from human-dominated may become difficult as virtually every urban green space is under some kind of control from human beings: even areas without control, have control, prohibiting control.[4] This notwithstanding a set of circumstances, where control on paper does not translate to control in

FIGURE 3.1 Counter-utilitarian biological growth within old city fabric of Bengaluru: top – ferns growing from the ventilator shafts of a bathroom pipes; bottom – small plants growing on a *chejja* (overhang), blocking the rainwater drainage pipe. July 24, 2015.

real time. For instance, a private yard may go unmonitored by its landowner where wildflowers may have blossomed outside of human care, or roadsides under control of the public authority may be subject to neglect in maintenance leading to wild growth, and so on. True, if the rightful authority so desired they would have the legal right to intervene – but clearly this legal right has not manifested into action in the instances above. Thus, forces beyond human control have resulted in a void within space and time, that temporarily creates pockets without human care. It is within these points across time and space that wild nature finds the opportunity to regain control of human-dominated systems.

Sites of convenience

Unlike human-designed landscapes, which respond to certain needs or purposes, spontaneous nature, despite its potential aesthetic, is not related to utility; and, as Gandy (2016) notes, they are often counter-utilitarian. For instance, a ventilator shaft that was designed for stale air to escape from bathrooms, is now blocked by growth of ferns (site A in Figure 3.1), leading to an unhealthy habitable condition. As another example, unplanned plant growth on *chejjas* (overhangs) and rooftops has ended up blocking the mouths of pipes that were meant for draining rain water (site B in Figure 3.1), thereby trapping moisture and potentially causing water intrusion into the structure. These settings are potentially harmful to the building in the long run, and, therefore, compromises the living environment of the inhabitants. Thus, the logic of utility does not always work to explain the existence of plant-built assemblages – for logically, they should have been removed. It demands a set of new logics to understand modes of production of these assemblages. As it would seem, nature finds ways into human-dominated systems through means which are more associated with sites of convenience.[5]

In plant sociology, a stone with moss growing on it would be called a *settlement*, as in my case would be a tree on a wall. However, by "sites of convenience" I do not just refer to physical settings – but the broad social, economic, and political context of the places. Because in cities, it is only convenient for nature to grow where the aforementioned factors are favourable, in addition to the geophysical and ecological factors. Landscape fractions in cities, at least those which I discuss in this book, have come up by chance, and not by human intention. Plant communities can be seen "adjusting" in an otherwise human-built landscape designed mainly, if not solely, for human purposes. So, it is safe to say – plants have grown wherever conditions were suitable, and more so: where conditions were enabling.

Convenience is a setting that goes beyond the physical elements of a soil substrate, nutrients, light, and air – what Swiss botanist Braun-Blanquet (1932) in his book on *Plant Sociology: The Study of Plant Communities* termed as "life conditions of a given habitat" – into incidental availability of alternative conditions in absence of some of the expected life conditions (p. 9). Water, for instance, in an urban setting is in short supply. Nature has to compete for water with humans in many cities where the water table has fallen hundreds of feet below ground level because residents have little or no restrictions for drawing up groundwater – leading to overdraft. The increasing temperature due to climate change and heat island effect instigated by reflective materials used in the construction of cities also affects moisture in an urban environment – increasing the rate of evapotranspiration in plants by as much as ten per cent (Zipper, Schatz, Kucharik, & Loheide II, 2017). These conditions make it difficult for plants to survive in cities. Not surprisingly, I noticed that grasses grew only in one patch next to a house where the resident unconsciously helped maintain soil moisture by emptying his bucket of water after washing his car every day. It was as though only the lack of soil moisture was preventing the rest of the street from becoming a habitat. This isn't a one-off encounter: overflow of water tanks, leaking pipes, and drains all add to soil moisture to help plant communities to emerge spontaneously and thrive. On this note, it becomes necessary to revisit Foucault's lines which capture the essence of convenience:

> *In the vast syntax of the world, the different beings adjust themselves to one another; the plant communicates with the animal, the earth with the sea, man with everything around him.*
>
> *(Foucault, 2005, p. 20)*

The convenience I bring out only concerns plant communities growing in the microsites, and this concept does not appear transferrable to humans or even nature as a whole. Braun-Blanquet (1932) notes that burnt patches of land have often offered the perfect conditions for a rapid succession of certain short-lived plant communities. Very conveniently, in the city of Bengaluru, there has emerged a tendency to burn leaves and litter within the neighbourhoods rather than taking them to an incinerator or a composting unit. While the emission from burning leaf litter is certainly counterproductive for human health and the global environment, it has facilitated new landscapes which are small-scale, and at times, resetting the wild spaces invaded with exotic species to offer another opportunity for local species to return.

So far, assemblages I have spoken of are constructs of natural growth and built environment. However, the existence of landscape fractions is influenced by a host of other anthropogenic and natural forces; which also form inseparable parts of the assemblage. These influences offer wild spaces the negotiating power to thrive in cities. Understanding such influences should explain why growth does not only happen in degraded urban environments but also pockets within affluent and economically well-performing neighbourhoods. More specifically, adding these influences into our framework of analysis should help us arrive at a complete picture that takes into account the context of different landscape fractions. In the following section, I will delve a bit more into the exogenous processes that help create the necessary conditions for small-scale botanical growth to come into being as well as continue to thrive inside human-dominated environments.

Social, economic, political, and natural processes

Processes that lead to the creation and existence of wild spaces in cities can no longer be attributed separately to roles of human and non-human actors – as traditional sciences used to. Bruno Latour characterises the role of non-human actors similar to that of human actors; based on their agency or power to act, and diverse configurations in which they occur (Latour, 2013). The plethora of demographic and political influences – including disruptions in metabolism and circulation – that a city is subject to, become part of the production process of these landscapes. Swyngedouw (2006) elucidates with examples from Jakarta to show how global dynamics can finally "[fuse with] local power struggles and socio-ecological conditions" and translate to consequences that allow the production of wild spaces in cities (p. 105):

Example 1: Disruption in financial flows

> *In the summer of 1998, the Southeast Asian financial bubble imploded. Global capital moved spasmodically from place to place,*

> *leaving cities like Jakarta with a social and physical wasteland where dozens of unfinished skyscrapers are dotted over the landscape while thousands of unemployed children, women, and men roam the streets in search of survival.*
>
> *(Swyngedouw, 2006, p. 105)*

Example 2: Disruption in climate

> *El Ninõ's global dynamics were wreaking havoc in the region with its climatic disturbances. Puddles of stagnant water in the defunct skyscrapers that had once promised continuing capital accumulation for Indonesia became breeding grounds and ecological niches for mosquitoes. Malaria and Dengue fever suddenly joined with unemployment, and social and political mayhem in shaping Jakarta's cityscape.*
>
> *(Swyngedouw, 2006, p. 105)*

Similarly, landscape fractions of spontaneous nature also exist from a combination of macro- and micro-level urban dynamics – insights that are discussed far less in mainstream academia. The decline of economic activities in a district means difficulty in arranging finance for the upkeep of landscapes and buildings. Such economic decay of urban fabric is seen in Plaza Republica (Figure 3.2) in Quito where an empty planter originally designed to accommodate ornamental hedges, is now reclaimed by spontaneous grass. This place is located at the heart of the business district of the city, and yet the plaza is deserted. Such plazas in cities are usually expected to thrive with shops, street activities, and outdoor seating. The absence of people, together with signs of neglect in terms of maintenance, explains the need for revitalisation and addressing the urban decay of this site. The graffiti on the walls, however, are traces of activities – albeit by those who perform their acts in the darkness of night.. Together, the human activities as well as natural forces, which are illegitimate, seem to have created a sense of place here. This could not have been accommodated if there was not an extreme economic disempowerment of the site. In theory, the existence of such landscapes can easily be erased with even the basic effort of vigilance and maintenance from public authorities. But to achieve overall conditions that allow such interventions often becomes difficult in reality – permitting wild nature to thrive.

FIGURE 3.2 Wild grass growing in empty planter boxes, and graffiti on walls of a structure in Plaza Republica of Quito. October 16, 2016.

Demographic changes such as migration and household size are important factors linked to the growth of spontaneous nature in cities. Coupled with the built heritage of Kolkata and Bengaluru, demographic change forms the necessary combination of forces that allow landscape fractions to be born, and thereafter continue to thrive. Old mansions in Kolkata can be seen to have biological growth peeking from their rooftops, or adding floral decorations to their ornate columns. Weathering of built surfaces in a line of buildings near Kumartuli, has exposed the intricate architectural and engineering details that went into the masonry and slab casting, presumably both when it was constructed during the British colonial period, and later when it was renovated (Figure 3.3). A variety of plants have found their home in various niches of this co-created environment.

Typically, in cities, there is a problem of division of plots that emerges from a sedentary tendency of generations of owners. But Kolkata's problem is somewhat unique – some residents owning the old palatial[6] mansions had enough determination to become mobile. Outmigration to

FIGURE 3.3 Plant biodiversity in a co-created environment: a variety of wild plants growing from weathered surfaces of an abandoned building in Kolkata. The disintegration of façade has exposed wooden beams and brick masonry from colonial period as well as steel reinforcement and concrete from more recent renovations. July 6, 2018.

other cities within India, or even abroad, has not been uncommon among the more recent generation of Kolkata's working population. This was perhaps necessary especially after the late 1970s when the economy of the old city began to exhibit stagnation. In the decades to follow, a lot of residents from the traditional areas have also moved to newer extensions of the city – such as Salt Lake City built in 1970s and now the New Town (near Rajarhat). Consequently, without stewardship, the buildings in the old city now remain uncared for. This has allowed for botanical growth to occur alongside the disintegration of façades which together create unique sculptures, like that of Kumartuli's buildings.

Closely related to the phenomenon of disintegrating urban fabric due to outmigration, is the general economics of the household and city – that allow assemblages of buildings and plants to thrive. The cost of maintenance of large colonial buildings is expensive. Not just in terms of repairs and restoration work that often involve old-fashioned and disappearing skills from service providers, but also because such buildings are highly inefficient when it comes to modern demands of cooling, heating, and general air conditioning. Coupled with falling income, the cost of repairing leaking pipes and cracking walls is something that families either ignore or avoid, causing a delay. On a city scale, both Kolkata and Bengaluru lack a reliable tree care industry. Wild growth such as fig and banyan growing on buildings need more than nursery trained professionals. These stranglers once grown can fuse with the gaps in masonry so intricately, that they may become extremely hard to separate without the risk of damaging the building itself. The skill level of professionals can easily be traced through the lack of heavy-duty technology that they use for arboriculture practices. Use of the latest technology, such as chainsaws and chipping machines, not only needs licencing for operation but also needs skilled professionals who can handle the equipment. Much like any other industry, tree care businesses also require resources to be invested, and a market demand that – as it appears for the cities – is yet to present itself as a business opportunity.

Natural forces are very important, but those plants have thrived particularly well (e.g., strangler fig), which can draw water from masonry and nutrients from the air. Often, plants that are parasitic have emerged the most victorious as they have alternative means to draw nutrients until their roots reach the ground. The local climate of the area matters as I have highlighted with examples of vegetation within the factories of Kolkata and Bengaluru in Chapter 1 on Morphology. But another aspect that plays an important role in the making of these unique assemblages are the materials with which buildings are built. They become substrates for plants to grow on. Brick which easily weathers offer a softer, more porous substrate than the quarry stones of Bengaluru. Incidentally, Kolkata is famous for its monsoon season where *kal baishakhi* storms bring heavy rainfall not to mention violent winds. Bengaluru's weather is timid in comparison: nominal rains, wind, and heat just enough to call a seasonal change. One would think that this means that Kolkata's buildings – which are made of lime and mud bricks – are easy targets for spontaneous nature. Indeed, they are, but that is not just a product of substrates and nature but a combination of socio-economic factors which have allowed far too much vegetation to grow on buildings in the core of the city.

Not all plants that grow on buildings are stranglers. Another medium of plant growth on buildings are *film-soils*.[7] By film-soils, I refer to thin layers of soil matter that has formed sediment. I first noticed plant communities growing on soil formation in thin layers while writing my book on urban ecology (see Deb, 2017); and later extended this observation using the phrase "film-soils" in one of my brief online articles for Soil Care Network – a facility at The University of Sheffield dedicated to research on soils founded by Leverhulme Fellow Anna Krzywoszynska.[8] I have the privilege of being a member of this network. In my article, I describe film-soils through the following narration:

> *Biological growth can still occur on the surfaces of buildings such as walls, cornices and roofs – without coming in contact with the earth's crust. Such growth can originate from very low-depth of soil-matter or even thin layers of dirt deposited on built structures. I refer to this growth-medium on buildings as "Film-Soils" … If we return to the most basic definition of soil, as supporter of plant life – one can find ample potential with Film-Soils on buildings to render services similar to other soils.*
>
> (Deb, 2018)

The KR market of Bengaluru has several footbridges that connect different wings of its colossal building complex. Here, I found a site that appeared from afar to be a rooftop garden[9], but in reality, had grown out of nutrient depositions gathered over one of the open-air footbridges which had been unused and locked up for a long time (Figure 3.4). The footbridge – somewhat in the centre of the building complex – is visible from multiple levels of the market premises; thus, the lush green shrub-growth created a visual relief in the centre for the thousands of people who visit different parts of the complex every day. Typically, such terraces have pipes for draining rainwater, but over time leaf-litter, dust, and other debris pile up – sometimes carried from the ground up by the wind, other times disposed from a higher rooftop by human agency – that leads to clogging of the rainwater drains. This leads to retention of high levels of moisture or even water pooling on rooftops, that creates the conditions for growth of moss and algae. Notably, species such as moss, that grow and require only a layer of dirt or just damp soil, as bryophytes can generally draw their nutrients from the air, water, or soil. As time passes, and with completion of the life cycle of these plants,

FIGURE 3.4 Plant growth on film soils: wild vegetation on locked footbridge of KR market building in Bengaluru, creating an appeal of a rooftop garden. July 24, 2015.

their detritus creates a thicker soil medium for larger plants to grow; this may lead to formations such as beds of grass or even wildflowers depending on seeds that human and non-human agencies have unintentionally transported onto the roofs. The market building is an apt example of how thin layers of soil can foster dense vegetation over time; however, this is not a one-off occurrence. I found such landscapes cropping up just about anywhere in the city where buildings are abandoned, or rooftops are neglected.

Natural forces are complemented by curtailed human capacities. Commuting on the overhead monorail of Bangalore, I could see entire stretches of rooftops that were so densely vegetated with tall grass, one could easily mistake them for fields. These could on one hand be exemplifications of my concept of film soils. On the other, it highlights a common fate of those rooftops of buildings that had remained locked or neglected. Imagine an office building or an industry where there is little reason for occupants to access rooftops; such a circumstance ends up creating blind spots in maintenance procedures. With very little reason for the staff or occupants to take notice of grasses that begin to grow on the building, nature finds a safe haven to thrive. The grasses dry up in summer and return during the rainy season. Legal disputes or concerns are another reason buildings remain locked. During this time, nature gets an opportunity to grow and thrive.

Barriers to access are not only physical but at times psychological – such as the electric substation set up within a factory in Bengaluru. Though this space is easily accessible, a danger sign prevents people from regularly accessing it without permission – of course for the purpose of their own safety. Other times, people themselves may be deterred from access to sites such as plants growing around electric poles or transmitters. All of this amounts to human incapacity shaped by risks, legislation, and rules.

The city is a political space, and nature finds its way to participate through a variety of channels. Conservation efforts in cities from Germany and Japan, as mentioned before in Chapter 1 on Morphology, have helped protect urban biotopes of scientific interests through planning tools and legislation. This inclusion in the legislation process has given a political voice to wild landscapes in the creation of urban space. Constructed in the late 1700s, Kolkata's colonial prison buildings are potential sites for such scientific excursions in the interest of biodiversity conservation. A rather recent political movement to protect Kolkata's old buildings has emerged through a series of appeals, protests, and other forms of public demonstrations by citizens who feel for these sites (Chattoraj, 2019). These are important events, for, sometimes, conservation ordinances of buildings with heritage value that carry biological growth, can in turn offer a safe haven for biodiversity. But of course, such plant-built assemblages, if not specifically identified as objects of protection, will be threatened if a decision is made for the buildings to be restored.

Rarely, but certainly, rules related to conservation are applied to urban-nature assemblages – such as in the city of Hong Kong. The stone-wall trees or Chinese Banyans (*Ficus microcarpa*) growing on the heritage walls around the city have now become part of the structure. A study for the Civil Engineering and Development Department of the Government of Hong Kong stated that in some cases these trees now provide additional reinforcement and stability to the structures as opposed to damaging them (C.M. Wong & Associates Limited, 2011). It follows that removing the trees would mean at least some amount of destabilisation of the stone structures. However, a subsequent study of selected instances across the city noted that stone-wall tree failures (due to storms) have not caused damage to the stone walls themselves (Jim, 2012). More recently assessing public attitudes, Lo and Jim (2015) explicitly reported that the mature trees combined with historic masonry made it a rare and appreciated setting among the local communities. Thus, the unique human-nature assemblage has a strong case to be protected in order to conserve architectural heritage and collective memory values associated with the site.

The occurrences of landscape fractions may also be for an extremely short duration – spanning anywhere between a few days to weeks. Ephemeral growth or flash landscapes can occur due to poor drainage of a site. I observed how a vacant plot used for parking by heavy bulldozers and trucks has been unintentionally transformed over time, creating the perfect conditions for water to stagnate and foster a landscape that depicts a miniature version of a vegetated lakeside (Figure 3.5). Comparing the evolution of this landscape over different periods of rainfall, I noticed the tyre marks below the clear water that cut deep into the soil. This sign informs about the repeated transgression of heavyweight automobiles that are responsible for creating depressions in the soil. The depressions hold the rainwater instead of channelling it to the storm water drains that perimeter the site. However, the reason I refer to this

FIGURE 3.5 Unintentional forces transforming land: deep cuts in soil created by movement of vehicles have retained water after rain, to create a flash landscape within a vacant lot in Bengaluru.

site as a flash landscape is because one can only enjoy the water for a few days before it disappears. As the water dries, so do the plants. Without the dramatic reflections of standing water, perhaps the assemblage would be incomplete; and without freshness of the grass blades, there might be no aesthetic.

Plant communities and their shared identity

Wild growth in cities seem to form in the most unusual places; however, there are many sites of urban wilderness across the city with the same combination of floral diversity, therefore making these sites incomparable to some extent in terms of botanical growth. They are difficult to compare because plants grow in so many different shapes and sizes, that organic growth becomes a common character, reducing the basis for comparison. In this book, the reader would, by now, already resonate with the natural palette being repetitive of ferns, figs, moss, and wildflowers as general categories. The exact species of course, as noted in Chapter 1 on Morphology, depends on the biophysical legacies of the region (Roman, et al., 2018). But more broadly, this phenomenon of grouping follows the property of association among plants where certain species are comfortable in growing around other species (Braun-Blanquet, 1932). The sandalwood tree, for instance, is parasitic as it needs to absorb nutrients and water from the roots of surrounding plants for it to grow. Thus, foresters at the soap factory in Bengaluru encourage spontaneous ground cover around the sandalwood trees within its campus.

One small patch of such wild growth, inside the soap factory, caught my interest. Here, I found a wide diversity of plants around a cast-iron staircase, competing for space and creating an interesting ensemble of plant community (Figure 3.6). However, the same groups of plants can be seen to grow all across the city – making this ensemble of plants quite common. The association of plants is abundant in the city making an otherwise interesting assemblage mundane. Perhaps what helps notice it more as a unique arrangement, is the architecture in which it is situated. Here, the stone masonry and the cast-iron staircase give signature and identity to the plant growth that help the observer assume the context of industrial surroundings. The architecture in this case integrates the ordinary association of plants on and around it, and makes them a part of an industrial legacy. Similarly, considering the opposite, certain species of rare plants growing on buildings can elevate the site identity to that of scientific importance. Identity is indeed more complex than aesthetics – they can appeal to social, cultural, economic, or scientific heritage. Communities are made through shared identities – built structures become reference points and other times the reference point is nature – the role shifts depending on the arrangement.

Despite the unique potential, so far, no tools have been devised to look at assemblages of built and natural. Consequently, cities delineate the sites of built heritage and natural heritage as separate entities that typically do not overlap with each other – e.g., a religious building versus a protected forest. This arguably reductionist approach in the delineation process only offer two extreme ends in the policy and planning of a city. For instance, if a site is considered built heritage, regulations are structured to protect the architecture, which often involves directives to control biological growth. And if the site selected is a natural heritage site, there is usually a blanket restriction from building more near it, assuming an anthropogenic encroachment would threaten the sanctity of the natural landscape. In contrast, recognising the

FIGURE 3.6 Iron cast stairs and stone masonry create identity for an otherwise common association of plants inside Sandal Soap Factory campus of Bengaluru. January 14, 2020.

classification of landscape fraction can provide a framework to look at biological growth on built surfaces and structures as a fused sculpture, which allows more options for an urban planner, an urban designer, or a conservationist. To that end, the idea of landscape fraction provides a framework for evaluation and selection by making each such site rare and unique; where the built and natural elements together become the unit of comparison.

Permanence of spontaneous settings

Just like a beard on a man's face or the tail of a lizard, plants too have the tendency to grow back in the same place after being cut down. Landscape fractions may be incidental, temporal, or even seasonal; but are certainly not a one-time phenomenon. Plants and shrubs show a high level of fidelity even after being subject to the social, political, and economic fluctuations of urban conditions. This is why sites of spontaneous growth have a tendency to fall in and out of shape and aesthetics with time; and there is in a sense, a reoccurrence – however irregular these cycles may be – or regrowth of nature, after plant growths have been removed or have died out because of unfavourable conditions. While social revolutions and economic restructuring can change the city at large, small interstitial spaces that harbour landscape fractions manage to survive.

Studying the patch of vegetation around the control box in Bengaluru for three years, it was clear that despite a harsh biochemical and socio-economic environment, the characteristic of site-fidelity holds good for even small-scale plant growth in urban areas. But it must be noted, that while the same plant species keep sprouting again and again, after being removed or after dying, their regrowth did not follow a seasonal pattern. But, instead, they seem to have followed the associative property of plants, where certain species grow because of the presence of other species. While the existence of natural components of this landscape has become unpredictable, the control box continues to stand as a marker of memory bringing unity in time and space. The built structure creates continuity in time: bridging the periods of when nature is scarce and when it is in full bloom. Resembling what Foucault refers to as *conventia* (Foucault, 2005), the proximity of natural to the built presents, to the observer, a sense of place for the plants. It points to one's senses how plants and human-made elements together define the existence of the novel arrangements I term as landscape fractions.

The above notwithstanding, there are also factors that make some assemblages be more transient than others; however predictable. By transience I am referring to an irreversible change in the site, that either removes the character of built or plant growth or stop their original cycles of reoccurrence. These factors may be environmental, such as growth of a new plant species through ecological succession that change the visual character and connotations associated with a space e.g., mature tree growth taking over early succession species of shrubs and ivy. Or the built fabric can undergo severe and practically irreversible changes. For instance, naturally over time, buildings lose their structural integrity to dilapidation and collapse. Other times, changes in the built elements of assemblages emerge from human preferences; which are predictable. It is unlikely that the local government in Bengaluru will decide to remove the control box any time soon; as that would impact the larger network of the public water supply system it is part of. In

FIGURE 3.7 Growth-cycle of streetside nature: Over three-years, plant associations disappeared and re-emerged around a control box in Bengaluru. During this time, the site has survived multiple human interventions in the area.

comparison, the government is more likely to order paving decisions or changes in road designs that can be experimented with, without severe impacts to the broader road network it is a part of.

Sometimes, spontaneous landscapes found in a city are products of incidental actions of completely unrelated human agencies who were simply acting independently, yet adjacent to one another. Take for instance, the example of a tree standing in front of blue metal panels that are set up temporarily as fencing, by a neighbouring construction site, to contain dust (Figure 3.8). In this arrangement, the blue panels offer visual focus and background to the tree. The tree forms the central element here – it has always been there, marking the edge of the property. Over time, other wild shrub growth has emerged all around the tree, due to neglect. A thin trail amid the wild growth in this landscape suggests repeated movement of pedestrians, and might mislead one to presume that this scene is set in a public realm. But in reality, the panels and the tree both belong to different private land parcels that belong to

FIGURE 3.8 Dependency of landscape fraction: the tree and blue panels belong to two sites within Bengaluru, with different ownerships.

different landowners. If either one of the owners decide to make changes to their own properties, the aesthetics of this assemblage would cease to exist in the same form. Such landscapes associate a higher degree of uncertainty in their existence, simply because there are more channels to disrupt the assemblage.

Meanwhile, on the same site, I also noticed a large curvy tree trunk that was probably felled from a tree nearby; it was left in front of the blue panels, likely to be chopped up as firewood or to be disposed of later. The wild plants should have grown all around the tree trunk, making it not easy to access. But spill-over debris – comprised of rocks and sand – from the adjacent construction site had made a neat clearing amid the wild grasses and shrubs. Placed within this clearing, the felled tree trunk now served as a bench for the residents to sit and enjoy nature. This is an example of non-design that has been adopted as part of everyday life by humans; demonstrating how assemblages can establish social links over time.

Notes

1 In Seoul (South Korea), an elevated highway was demolished to uncover and revitalise the Cheonggyecheon stream (Lee, 2006). In Raipur (India) about 70 dilapidated public buildings in a 19-acre site were demolished by the local government to make way for urban forestry and greening (Sivaswamy, 2017). In Faridabad (India) illegal structures were demolished to make way for urban forestry (Roy, 2021).

2 R. Carrington. Management of urban nature in UK cities, interview with A. Deb, October 9, 2022.

3 In mapping the spatial structure to develop a more grounded classification of urban green infrastructure using high-resolution satellite imagery, Dennis, et al. (2018) refer to urban systems as human-dominated systems. The authors assign dominant traits of neighbourhoods through green spaces such as parks, gardens, tree-lined neighbourhoods, farms, and woodlands that are heavily subject to human management.

4 I remain grateful to Prof. Bruno Latour for directing my thoughts towards this enquiry.

5 I partially borrow this concept of convenience from Foucault's proposition of adjacency as a sign of relationship rather than merely physical proximity. Foucault (2005) writes, "*We see mosses growing on the outsides of shells, plants in the antlers of stags, a sort of grass on the faces of men; and the strange zoophyte, by mingling together the properties that make it similar to the plants as well as to the animals, also juxtaposes them. All so many signs of 'convenience'*" (pp. 20–21).

6 Kolkata has been referred to as city of palaces owing to the proliferation of garden houses (or "*bagan bari*" in Bengali) that reflected the power of the wealthy and pristine European enclaves (see Sen, 2017, pp. 54–59).

7 For a related term, "bio-film", see Whitby pier discussed in Chapter 4 on Aesthetics and treatment tank in the Sandal Soap Factory in Chapter 6 on Co-production.
8 See the website of the Soil Care Network to learn more: www.soilcarenet work.com.
9 The reminder of a "rooftop garden" was a common response from discussants who were shown photographs of similar settings. A. Kumar and N. Luke, photo-elicitation of landscape fractions with A. Deb, Bengaluru, June 21, 2019.

References

Braun-Blanquet, J. (1932). *Plant Sociology: The Study of Plant Communities*. (G. D. Fuller, & H. S. Conrad, Eds.) London: McGraw-Hill Book Company, Inc.

C.M. Wong & Associates Limited. (2011). *Study on Masonry Walls with Trees*. Geotechnical Engineering Office; Civil Engineering and Development Department. The Government of the Hong Kong Special Administrative Region.

Chattoraj, S. (2019, May 11). *PIL against razing of heritage jail buildings*. Retrieved May 14, 2021, from *The Times of India*: https://timesofindia.india times.com/city/kolkata/pil-against-razing-of-heritage-jail-bldgs/articleshow/ 69274953.cms.

Deb, A. (2017). Understanding urban 'wildlife'. In A. Deb, *Introducing Wildlife in Urban Ecosystems* (pp. 13–43). New Delhi: Copal Publishing Group.

Deb, A. (2018, December 21). *Film-Soils, There is a Layer above the Crust*. Retrieved June 8, 2022, from Soil Care Network: https://www.soilcarenet work.com/single-post/2018/12/21/Film-Soils-There-is-a-Layer-above-the-Crust.

Dennis, M., Barlow, D., Cavan, G., Cook, P. A., Gilchrist, A., Handley, J., … Lindley, S. (2018). Mapping Urban Green Infrastructure: A Novel Landscape-Based Approach to Incorporating Land Use and Land Cover in the Mapping of Human-Dominated Systems. *Land*, 7(1), 17. doi:10.3390/land7010017.

Foucault, M. (2005). *Order of Things* (Reprint ed.). New York: Routledge.

Gandy, M. (2016). Unintentional landscapes. *Landscape Research*, 41(4), 433–440. doi:10.1080/01426397.2016.1156069.

Gandy, M. (2018). Cities in deep time: bio-diversity, metabolic rift, and the urban question. *Cities*, 22(1), 96–105. doi:https://doi.org/10.1080/13604813.2018.1434289.

Jim, C. Y. (2012). *Study on Stonewall Trees: Maintenance Approach for the Six Stonewall Trees on Slope no. 11SW-A/R577, Bonham Road*. Highways Department. Government of the HKSAR. Retrieved May 22, 2021, from http://www.districtcouncils.gov.hk/central/doc/2012_2015/en/dc_meetings_doc/ 9351/20150827_File1_ReportText_WallTreeSurveyText_20130108R2.pdf.

Latour, B. (2013). Biography of an inquiry: on a book about modes of existence. *Social Studies of Science*, 43(2), 287–301.

Lee, I.-K. (2006, March 15). *Cheong Gye Cheon Restoration Project*. Seoul. Retrieved December 30, 2022, from https://www.seoulsolution.kr/sites/default/

files/policy/%5BEN%5DCheong%20Gye%20Cheon%20Restoration%20Project.pdf.

Lo, A. Y., & Jim, C. Y. (2015). Community attachment and resident attitude toward old masonry walls and associated trees in urban Hong Kong. *Cities*, 42 (A), 130–141. doi:10.1016/j.cities.2014.09.006.

Roman, L. A., Pearsall, H., Eisenman, T. S., Conway, T. M., Fahey, R. T., Landry, S., … Jenerette, G. D. (2018). Human and biophysical legacies shape contemporary urban forests: a literature synthesis. *Urban Forestry & Urban Greening*, 31, 157–158.

Roy, S. (2021, August 9). Following demolition, urban forest to come up on 150 acres in Khori village. Retrieved December 30, 2022, from *The Hindustan Times*: https://www.hindustantimes.com/cities/gurugram-news/following-demolition-urban-forest-to-come-up-on-150-acres-in-khori-village-101628530461591.html.

Sen, S. (2017). Colonizing Kolkata: from a city of huts to a city of palaces. In *Colonizing, Decolonizing, and Globalizing Kolkata: From a Colonial to a Post-Marxist City* (pp. 37–76). Amsterdam: Amsterdam University Press.

Sivaswamy, R. (2017, June 15). *While the World Fells Trees to Construct Buildings, Raipur Demolishes Buildings to Make a Forest*. Retrieved December 30, 2022, from The Better India: https://www.thebetterindia.com/104935/raipur-oxyzone-natural-forest/.

Swyngedouw, E. (2006). Circulations and metabolisms: (hybrid) natures and (cyborg) cities. *Science as Culture*, 15(2), 105–121.

Zipper, S. C., Schatz, J., Kucharik, C. J., & Loheide II, S. P. (2017). Urban heat island-induced increases in evapotranspirative demand. *Geophysical Research Letters*, 44, 873–881. doi:10.1002/2016GL072190.

4

AESTHETICS

Unintentional landscapes

The traditional understanding of nature itself, is something as political economist John Stuart Mill noted to be "without the...intentional agency, of man" (Mill, 1874, p. 8). Not only that, here "man" would be a simplification of "human", but also, agency is duly qualified as intentional. Urban nature however, has been subjected to a tremendous amount of anthropogenic pressure. For city dwellers, landscape designs within the built environment are often the closest experiences of nature. These human-made arrangements of plants are thought to be artificial; or in other words "unnatural". Yet, Glenn Parsons, a contemporary philosopher of aesthetics, draws attention to how plenty of "nature" exists in the midst of "civilisations" (in our case, cities) (Parsons, 2008). Parsons refers to climatic dramas (e.g., rainfall) or animal behaviours (e.g., mating) that embody the essence of wilderness in them; albeit while maintaining, that natural phenomena are now well known to be influenced by human activities in some way or another. Notably these natural experiences do not have to be limited to spontaneous vegetation – for mating of frogs, foraging of birds and bees, snakes hunting rats, dew drops on leaves are all common sights in well-groomed backyards as well. Therefore, in the light of how various authors have interpreted the "unintentional" quality of nature, perhaps it is useful to bear in mind nature is not necessarily "landscape".

DOI: 10.4324/9781003374602-4

Owing to dictionary definition that relates to this book's subject, a *landscape* is "a large area of land, especially in relation to its appearance".[1] However, my object of study is only small-scale wild growth which is a subset of, but not all of the landscape, rather fragments within that appeal to aesthetics. This is expressed through the accompanying word *fraction* whose closest meaning is "a small part or amount of something".[2] A small amount of a landscape or a *landscape fraction* of spontaneous nature is a holistic assemblage of different elements – living and non-living – that have come to existence without the intentional intervention of human beings. The human agency may well be present, but without the goal of fostering growth of such nature or creation of such landscapes. To affirm the premise of such unintended assemblages, I revisit Matthew Gandy's description of "unintentional landscapes":

> *An unintentional landscape can be defined as an aesthetic encounter with nature that has not been purposively created. It is a space that has nonetheless undergone some form of sensory enframement, perhaps only momentarily, as a focus of attention. It is not necessarily a space associated with visual delight or even disorientation, in a narrowly aesthetic sense, but something much less easy to categorise or define.*
>
> *(Gandy, 2016, p. 434)*

Wild plants growing in small marginal spaces of the city can be visually intriguing –elevating the status of other ordinary nature as landscapes. Landscape fractions that are formed through small-scale natural growth within cities are also framed in sensory experiences of spontaneous nature, which as Gandy defines – may be momentary. Interestingly, when an assemblage renders "visual delight", it perhaps at least momentarily overrides in our minds the negative connotations of spaces such as damage to structures, urban decay, species invasion, allergies. Gandy notes that spaces where plant growth is "counter-utilitarian" can also become aesthetic, although maintaining that aesthetic is not just visual appreciation but is entangled in our overall sensory experience (Gandy, 2016). I discuss examples of counter-utilitarian growth in Chapter 2 on Existence. While enquiring about aesthetics of abandoned spaces, Gandy's references are to those spaces that are far larger than mine – grasslands in an airport, vacant lots, and other interstitial spaces which are either vestigial, residual, or are perceptibly under transition. But the underlying complexity of sensory experiences remain broadly the same; they include themes such as sense of place, abandonment, memory, emotions and representation of landscape, modern ruin landscapes and abandoned sites, aesthetics of landscape, materiality of values, and so on. In addition to the discussions in this chapter, I also touch upon some of these aspects later in Chapter 6 on Co-production.

Post-industrial landscapes often have meant abandoned mines[3] and other industrial establishments that are often large tracts of land outside of the city. Of course, similar sites exist within cities as well. Furthermore, objects of my interests are even smaller, since microsites are often created with fragments of such landscapes. Often in cities landscapes embody a more packed urban built form where potential spaces of accidental botanical growth that are often a failure of intended design. Take the example of a day care centre in Bengaluru, abutting a main road (Figure 4.1): given the vantage point of the property, the outer walls of this day care centre have become a useful surface for the day care's own advertisement to the passer-by. It is not uncommon for sites to develop incrementally,[4] and in this case, perhaps an obvious sign is the extremely narrow recess along the road – which could have been better integrated into the property, had it been planned and built at once. While it is not known to me why the recess actually exists, it is hard to imagine any day-care related utility in that little space of roughly 3m x 0.5m. Perhaps a conflict of ownership, perhaps an unrealised planting strip for beautification – whatever the reasons, now does not matter. The strip offers an oasis of green with wild plants that display a variety of accents. The plants combine with the colourful typography of the advertisement painted on the wall offering a rich visual experience to the people using the busy road.

In the above example of day care centre, while the architectural elements were rather unornamental, the colours of graffiti on the walls perhaps made up for the vital signs of human presence in the fraction. Other fractions of assemblages may have more pronounced architectural examples – I will discuss some of them later in this chapter. More broadly, appreciation of spontaneous landscapes that relies on sensory stimulation, suggesting that an observer's perception plays an important role in determining aesthetic value. One aspect guiding perception is imagination (Parsons, 2008). Novel ecosystems may lack natural analogues, but imaginations can relate plants growing on rooftops to grass

FIGURE 4.1 Unintended nature appropriating residual space in front of a day care centre in Bengaluru. The letterings on the walls are painted in different colours to attract passer-by attention.

fields, or derelict buildings as caves, and so on (Gandy, 2013). One can perhaps relate to hobbies of building animal shelters in home gardens, to understand how visual resemblance of sites to natural habitats might help more wildlife adapt to the urban environment.

This sensory appreciation has helped unintentional landscapes become a notable tool in building social capital. To some, the interesting, and at times, unusual ensembles of spontaneous nature, made them worth posting on their social media pages. This is in no way an insignificant quality of the landscapes, as aesthetics of spontaneous nature become evident when such assemblages are "Instagram-able".[5] When people find unintentional assemblages of nature interesting, they may click a photograph to record something that is desirable and yet an unusual and possibly momentary. The proliferation of cameras in our mobile phones, have especially assisted in the recording of these landscapes. Social networking websites and other image hosting Internet archives are where people would like to post pictures of such unintentional landscape publicly. Ultimately, when Internet users like or dislike these images, they are directly bringing together people who have been drawn to such assemblages. That is how assemblages of spontaneous nature, otherwise deemed as wasteland, enter the formal market of aesthetic valuation.

Chapter 2 on Meanings should offer the multiplicity of representations that unintentional landscapes under the ambit of my study can hold. But discussing aesthetics of such sites through images can be problematic, as not all visually appealing natural settings are considered desirable in terms of their utility; reflecting the duality landscapes hold in terms of its visual and physical properties. I discuss some of these counter utilitarian aspects in Chapter 3 on Existence. Environmental psychologist, Rachel Kaplan, points to the importance of visuals in more accurate articulation of aesthetics (Kaplan, 1975); yet without the physical understanding of wild landscapes, discussants may end up highlighting aspects that appeal to the eye, discounting some of the negative notions that surround the same fractions – say, in terms of root action, foul smell, plant borne-allergens, insect-borne diseases, and so on. To address this problem of looking beyond the visual, the book has separate chapters to understand cultural, economic, and ecological values (Chapter 5 on Collectives); as well as how socially constructed discourses affect perceptions of such nature (Chapter 7 on Frames). However, in this chapter, the focus will be on art values associated with the assemblages.

Geometry and spontaneity

About 50 metres from a busy metro rail station in Bengaluru, there lay an old bus stop, redundant in its use – its steel frame corroding, and paints washed away in rain. The backlit holder for the advertisement on the top displayed an outdated banner which was torn – and it is hard to tell if that happened due to a rough storm or vandalism by delinquents who threw rocks at it to entertain themselves as an escape from boredom and idleness. The rainwater pooled at the foot of this structure, taking advantage of the weathered drainage of the street side. The stagnating water has helped a bed of grass grow. The fresh green blades of the grass are a rather soothing contrast to the torn red banner. The small patch of plants has grown in the shape of a crescent along the kerbs. Nature here, has taken the shape allowed by the built fabric.

It is perhaps an important detail, that the formation of wild spaces is guided by pre-existing urban design, and is not entirely untouched. For instance, the traditional area of Bengaluru already has a compact urban fabric. Barring exceptions, buildings here should not be very old; yet many buildings would easily date back to half-a-century. Unlike examples from Kolkata, buildings that are part of Bengaluru's traditional fabric were also not built for grandeur or for monumental value, but for everyday use. An aged resident who grew up in Bengaluru may reminisce the old theatres, the racecourse, the traditional market area, and hotels that supported not only residential population, but also for visitors to whom this area around the transnational railway station was the first experience of the city. The built form has developed over time to house the needs of a global economy: there are now major bus terminals (both domestic and inter-provincial) and a metro interchange that is critical for the light rail network across the city.

The trees grow in fluid shapes, over buildings that have been designed by different architects, for different people, in different styles across different periods of time. In the practice of architecture, elements of nature are nothing new, as motifs and cornices of Greek, Roman, and Baroque architecture – or more recent works of art nouveau – all carried ornamental work that imitated natural elements.[6] In Kolkata, sites of landscape fractions are composed of colonial buildings which have naturally undergone wear and tear to create conditions on its surface for plant life from between eroded mortar and bricks. In Bengaluru and Quito, neglected postmodern buildings and spaces around them have allowed plant growth over advance concrete and steel structures. Colonial buildings typically have the charm of ornamental columns and motifs on walls, whereas the post-modernist architectures are less expressive in that sense, though can be equally impressive because of their simplicity.

Because spontaneous growth outside human care, lack a predictable geometry, describing the beauty of such landscapes can become hard. Although wider research has shown that people find naturalness of landscapes equally appealing not just in the natural areas but also within human dominated systems (Knez, Sang, Gunnarsson, & Hedblom, 2018; Krapez, Hughes, & Newsome, 2021). Such values sometimes lead architects to artificially create the perception of spontaneity in landscape designs of formal urban settings (e.g., golf courses, backyards, and public parks) for their clients who desire such values. This sense of appreciation is an important part in valuing landscapes for their "aesthetic preservation" (Parsons, 2008, p. 32). Parsons (2008) also cautions about appreciating aesthetics of nature without a framework as they may lead our thoughts "astray" (p. 33). At the same time, the author in his book *Aesthetics and Nature* highlights at least three approaches to appreciation:

- **Formalism:** based on visual elements (or art values) of geometry and colour; however, the criticism being that it disregards spontaneity of nature.
- **Science based:** how the body of knowledge influence our sense of aesthetics.
- **Pluralist:** a more cognitive approach that takes into considerations both arts and sciences, rendering the previous frameworks as inadequate.

Observers are of course entitled to adopt their own framework to achieve the common goal of appreciating nature.[7] Yet, the pluralist approach seems to offer the broadest base as framework; therefore, being more inclusive. The pluralist approach is prominent in Matthew Gandy's works as he connects the sciences and the arts in interpreting wild spaces in contexts otherwise, and often if not always, made synonymous with "wastelands" (Gandy, 2013). Gandy's paper *Marginalia: Aesthetics, Ecology, and Urban Wastelands* brings together the works of film makers, flower decorators, biologists, writers, and artists to show how imagination plays a significant role in determining the aesthetics of these sites. This encourages, at least in imagination, a shift towards an appreciation of wild spaces in cities, thereby hopefully refabricating their status as part of contemporary taste in aesthetics.[8]

However, the role of formalism in appreciation of landscape fractions remains to have a place; and perhaps serendipitously, the term *fraction* bears a close resemblance with the term *fractal*. Both terms are mathematical and are applicable – in the sense, a fraction represents proportion or a small part of a whole, while fractal associates with repetition in visual patterns. Both terms may also apply to natural and built components of the assemblages. Fractals are rhythms that exist in nature, such as petals of flowers, leaves of trees and trees themselves. Picture how the narrow-scattered leaves of shrubs from the day care can offer a striking contrast against the broad-leaved shrubs, that nevertheless come together to form a harmonious picture. Such repetitive floral patterns in general are considered to be aesthetically pleasing and stress reducing (Taylor, 2006); consequently, they have also been referred to as "biophilic patterns" (Robles, et al., 2021). However, researchers also note how the aesthetics of fractals are more universal in nature i.e., they are appealing to our eyes regardless of whether they are natural or human designed (Spehar, et al., 2003). Indeed, they may also be artificially created through repetitions in exposed masonry, motifs, and so on. The noticeably different patterns create separate foci, splitting the assemblage into smaller intriguing components, where geometry becomes more predictable.

Perhaps then, Benoit B. Mandelbrot's book, *The Fractal Geometry of Nature* published in 1982 warrants a noteworthy mention, even if statistical expressions do not occupy the centre stage of my analysis. The work delves into details of some of the irregularities in nature that break the more simplistic geometric patterns. Mandelbrot (1982) explores the logic of geometry in more challenging shapes of nature such as coastlines, clouds, mountains, and trees. This work delves into describing these irregularities in terms of statistics and explains how the fragmented patterns make sense in different scales. The work also reinforces how symmetry and repetition of mathematical patterns create complexity in aesthetics. And how absolutely vast landscapes that seem oddly shaped to start with, begins to find stability in their atomic or microscopic structure. Some of the patterns discussed in this book can be seen to still follow more elementary mathematical principals of aesthetics such as the *Fibonacci progression* and *golden ratio*.

Landscape fractions which are more specifically visually aesthetic precincts, are selective and demand a visual analysis. Analytic tools based on mathematics (such as the *golden proportion* or *rule of the thirds*) are not uncommon among artists and architects. Beyond these conventional means, there have been other methods to analyse images for their geometric beauty. One process can perhaps be related to what Gillian Rose in her book *Visual Methodologies: An Introduction to Interpretation of Visual Materials* explains as "the Good Eye"; where the author stresses the importance of contextual information in describing images i.e., to understand the site that is portrayed by the image (Rose, 2001, p. 33). The Good Eye relates more to explaining the composition of the objects in the images rather than the production of images. In this chapter, I follow a similar process in analysing photographs of aesthetic assemblages that I have collected across various cities, to arrive at a set of typologies. Perhaps worth noting, that understanding architecture through photographs can be different from that of two-dimensional art works, as it involves interpreting both external appearances as well as purpose of spaces. The production and interpretation of these photographs are explored more in Chapter 7 on Frames.

Three visual typologies of assemblages

Landscape fractions are composed of two broad sets of elements: one the built, the human-made or conceived; and another is the natural, the biological or the floral. By nature, I mean broadly plant growth. If one assumes solely the effect of visual upon the viewer or what Alois Reigl terms as *Stimmung*,[9] then, one can easily identify three typologies in which these compositions can be categorised: (i) the architecture heavy composition, where buildings are the dominant part of the attraction; (ii) nature heavy composition, in which plants play the lead role in drawing the attention of the viewer; and (iii) balanced elements, where nature and built are equally calling the viewer's attention. When one looks at these typologies with the principles of architectural design in mind, then the first two would relate to contrast and emphasis where the third would relate to balance and harmony.

If one flips over the pages of this monograph, the photographs would all serve as examples. In this chapter, I am going to highlight few selected landscapes, which I found typically representative of the three typologies aforementioned. These span from examples in Whitby,

Kolkata, and Quito. I describe these selected arrangements with principles of design in mind, to help my readers understand that beauty, in this case, is not only held by the individual elements of built and floral, but by effects produced in their combination.

The first typical arrangement is architecture heavy. On the east-coast of the UK, within the town of Whitby, a pier stands over the River Esk, next to the mouth where it meets the North Sea. The banks of the riverbank become active in the evenings, when virtually residents of all ages come down to participate in their most favourite activity: crabbing. In Whitby, crabbing is very much a family activity and families can be seen with more than one "crabbing kit". A crabbing kit includes a handline with a hook which baits the crabs, a drop net for trapping the crabs, and a bucket where crabs can be kept after they are caught. The pier serves as one of the many platforms near the river which allow citizens to select a good vantage point for catching crabs. It is not until after the tide recedes, is one able to see the algae growing on this pier (Figure 4.2). A set of stairs allow people to step-down into dense seaweed growth on the

FIGURE 4.2 Low-tide exposes seaweed growth at the walls and foot of a seaside pier in Whitby. June 3, 2017.

river bed. These biological growths are landscapes in themselves; however, when combined with the pier's masonry, they create a unique sense of enclosure. The pier's design, materials, and heritage play an important role in the aesthetics of this setting.

The second arrangement is nature heavy. Here the architectural elements play a far lesser role and the biological growth is more dominant. I choose a vacant lot in Kolkata enclosed from three sides by buildings with strong architectural character (Figure 4.3). Nature here dominates so much that, the arrangement can be, at least with a bit of imagination, compared to discovery of a site which belongs to an abandoned civilization in the middle of a forest. In reality, the site is tucked away amid a busy residential neighbourhood packed with large buildings which were built during the colonial period. Why the site is empty is anybody's guess – the remains of plinth from a structure which was existing makes it clear that a building existed here but has been demolished to make way for new construction. The compound is gated and locked to demarcate clear ownership of property. However, delay in sale and

FIGURE 4.3 Overgrown spontaneous vegetation within a vacant (p)lot in Kolkata, creating an impression of a forest. March 11, 2018.

prolonged neglect of the property has allowed flora to thrive long enough for the vegetation from the vacant site to have crept up the enclosure walls. The remains of the buildings are not the only human-made objects in this landscape. I noticed the giant black coloured PVC barrels lying in between the plant growth. Any local resident would identify these barrels as overhead water tanks, because they are widely used around the city. Despite the presence of large built elements, the rampant growth of ivy draws all the attention. I can relate to Parson's idea of wilderness as a momentary "escape from the city" (Parsons, 2008); as the overgrown site offered an impression of a forest, provoking imaginations that momentarily allowed my thoughts to drift to natural spaces far away from the city.

The third arrangement has a balanced role of both architectural and natural elements. A large abandoned building in the centre of Quito city forms the backdrop for another abandoned construction site in front of it; the two sites are separated by a wall. This concrete platform appears to be the unfinished flooring of a construction on hold which has wild grass growing over a concrete platform (Figure 4.4). Abandoned columns

FIGURE 4.4 Wild grass, construction site underway, and architectural ruins, together compose a balanced visual within Quito. October 16, 2016.

indicate that roofing was planned as part of the construction, though has not yet been realised. This site in the front abutting the road can now conveniently be used as parking space. Although the building in the backdrop physically dominates this scene, a light tint of beige belonging to the masonry respects the light green of the grass in front, together creating a harmonious composition that is soothing to the eye. Moreover, the large building runs linearly in the backdrop to unite the small and disjointed patches of grass growth and concrete flooring in the front. But between these two sites is a deep gorge separating the plot with the parking lot and the building behind. So, two different plots, which are adjacent but not connected, needed to have a completely unique set of forces that create convenience for this landscape to draw its beauty. A change in either one of these land parcels would mean a change in the aesthetic quality of the landscape.

Monumental values

Incidentally, both sites in Quito and Kolkata mentioned above, are derelict structures, but are able to move away from meanings of decay through their aesthetics to become artefacts. It would seem that the combination of built and natural has the potential to elevate everyday nature into iconic nature. But a monument is more than just a landmark, and perhaps a key distinction would be how the tangible features of assemblages associates with the intangible fabric of the society. In the previous chapters, I have already offered some examples of how memories can attach themselves to landscape fractions. I return to the arguments I made about era and age in Chapter 2 on Meanings. This sense of time is captured in the elements which compose landscape fractions, thus making them comparable to monuments. Monuments, according to Reigl, have embedded in them memories from the past on one hand, and on the other can also serve values of aesthetics and usefulness – which are relevant even in the present day. Kolkata's massive old mercantile buildings, now uninhabited or occupied by squatters show signs of decay, abutting the banks of Hooghly River (a tributary of the Ganges River network). The built spaces are a reminder of port activities that used to go on at a time during the reign of Nawab Siraj ud-Daulah and the arrival of the East India Company from Britain. However, I was unimpressed by the plant growth on some of these buildings, which seemed to reflect more of decay than age. The plants did not have enough character to add to the buildings. But as one walks through the city, the nature, quality, discourses, and scale of the spaces begins to change, and green begins to add age to the character – diminishing signs of decay.

Many scholars have explained the different dimensions of how collective memories are produced, and the role of urban regeneration in preserving them (Lak & Hakimian, 2019; Bélanger, 2002; Dunin-Woyseth, 1996). In case of landscape fractions, such memories can attach themselves separately to built and floral elements – and if these ensembles are not understood well, the city might lose historicity attached with plants during regeneration. Research and experiments in different parts of the world have brought out how some plants can hold collective memories of the city's past (Lo & Jim, 2015; Ryan, 2013), however historical accounts of plants are often less documented or propagated than those of iconic buildings. When histories of plants are not known, then the role of plants is limited to only adding the essence of age to buildings – and only the buildings decide which memories to be triggered for the viewer. Returning to Reigl's concept of pastness, virtually any plant growth can add age to structures – even devoid of particular memories. Having said that, particular species such as moss and lichens are more associated with heritage structures than, say, parthenium – the shrub is not looked at in terms of its exotic migration path from Latin America to Asia, but simply as a weed that can exist almost everywhere.

Memories are powerful ingredients of creating monuments, even after the use of structures have become obsolete. In Kolkata, along Amherst Street, sometime presumably in the early 1990s, an unusually large boundary wall measuring as high as 12 metres (36 feet) in height was erected within Hrishikesh Laha's residential property to preserve the privacy of the family from the activities of Raja Rammohan city college in the neighbouring plot (Figure 4.5). The boundary wall was erected merely as a visual barrier that has now become redundant in purpose, yet the associated story of the wall becomes a marker in time, that allows one to understand the importance someone would place on privacy – elevating its value as a cultural artefact. However, today the wall is vestigial, as buildings have not only come up all around the plot, but the residents seldom visit their property. Yet, as the city becomes more compact, the wall's message then becomes important as it helps understand the importance of non-intrusion in peoples' lives. But then what of the plant growth? The combination of figs and ivy intertwined with colonial brick masonry and its floral motifs create an aesthetic that is rare – bringing an otherwise boring yet ornamental structure to place it more firmly in a time that has passed. Kolkata is otherwise riddled with countless post-colonial buildings without plant growth – bearing an impression of the city's present rather than its past.

Plant growth reflects age and character for heritage buildings; but as with notions of decay, unmanaged plant growth may not always make buildings or places attractive. This vastly varies with context. When shown pictures of heritage buildings, one interviewee pointed out that despite the green cover the overall image was not always enhanced.[10] Some responses also asserted that certain structures looked bad without the green cover. There is certainly such a thing as "the right amount of green" since not having "much green"[11] also took away from the aesthetics of wild spaces. In tropical countries, green can be more easily perceived as good because it does not necessarily threaten urbanisation. But Doherty (2017) in his book "Paradoxes of Green Landscapes of a

FIGURE 4.5 Boundary wall erected as visual barrier for privacy in Kolkata, now has a host of wild plant growth to become an aesthetic, ecological, as well as, cultural artefact. July 14, 2018.

City-State" studies Bahrain in the desert region of the Middle East to explain why green is not always good for urban – greening demands resources, putting more pressure on natural resources than it would in the city's traditional context. Although this observation may not necessarily relate to unintended landscapes emerging outside human care, it at least leads to an understanding that the assemblages will be evaluated both in parts and as a whole while considering the entirety of the city context.

Nature which is integrated into the everyday activities of people will have more opportunities for being appreciated and absorbed into a community's collective memory. Matthew Gandy in his essay on unintentional landscapes mentions that unintended spontaneous nature is not always associated with visual delight, and their aesthetic properties as "narrow" and hard to describe (Gandy, 2016). Perhaps the word "narrow" can be expanded as a narrow window of opportunity which pops up during the momentary encounter for a viewer to appreciate the assemblage. But this very property of landscape fractions to incite an emotion within a fleeting moment enables a connection between people

FIGURE 4.6 Banyan tree next to the public water standpipe in Kolkata becomes everyday nature for people using the standpipe. July 9, 2018.

and nature. With increasing the incidence of visual interactions, this connection can be sustained. An example of such sustained encounter is the banyan tree growing next to a public water standpipe in Kolkata (Figure 4.6). The water standpipe is visited by people for their everyday needs, and thus the tree around it becomes part of the visitors' construct of everyday nature.

Post-industrial transformations

In recent decades many cities across the world have shifted out of the traditional manufacturing economy to encourage other businesses such as finance and information technology. It follows that people living in these cities also have a renewed preference of aesthetics now which reflects in their preference for glass-clad buildings like that of the Swiss Re Tower in London and decorative egg crate walls as in the case of Masdar Institute in Abu Dhabi. The typical iron-built structures which are depictive of the industrial era are increasingly becoming obsolete to give rise to an architectural language that is more representative of the present age – and what I am referring to as post-industrial. In post-industrial cities, sites of heavy industrial structures now only remain in the cities as a legacy. The soap factory in Bengaluru is a good example that can represent this transformation, although this space is partly operational, and is typically restricted from public usage today. Occasionally, the factory has offered guided tours to visitors to showcase the city's industrial legacy.

The Sandal Soap Factory in Bengaluru has a 34-acre campus which accommodates a variety of substrates. Sandalwood trees can cost up to INR 6000 (approx. USD 93.42) per kilogram (as per 2015 data and prices); thus, not surprisingly, the campus is highly guarded by walls and security personnel. Restricted for trespassers, wild nature within the factory campus is strikingly different from the rest of the city. Created by substrates of coal and other chemical effluents of the treatment plants, the factory had a peculiar variety of flora (Figure 4.7 and Figure 4.8). The soap factory is also one of the last remaining large-scale manufacturing sites in Bengaluru that run on coal within the inner-city area. The pitch-black colour of the coal yard that allows patches of grass to grow, is a remnant of the industrial era of the city and creates a unique setting within the city's landscape since, in Bengaluru, the soil otherwise is predominantly red. This is a rare sighting indeed as, in many parts of

FIGURE 4.7 Coal yard within the soap factory industrial site of Bengaluru. The black soil contrasts against the otherwise predominant red soil of the region. January 14, 2020.

the country, factories with boilers that run on coal have been relocated outside of the city to control emissions.

Other similar spaces in Germany have been transformed[12] into plazas and recreation parks by preserving the architectural legacies (see Vaughn, 2015). In her undergraduate thesis submitted to University of Arkansas in 2015, Vaughn argues how post-industrial landscapes have the potential to offer many benefits which are linked to human interaction and environmental psychology.[13] However, it is interesting to observe through Vaughn's documentation of Landschaftspark in Germany that while industrial structures have been retained and repurposed by retrofitting with climbing installations and parks, nature has been designed and not retained as spontaneous. Edensor (2005) discusses the uses of such industrial sites and explains how even in their fully abandoned and neglected existence, they can offer more playful and tactile experiences. Such projects demonstrate an understanding of the built elements of industrial heritage but not spontaneous nature.[14] In comparison, through the landscape fractions framework, I am suggesting to take a look at the combined assembly of built and plant growth these sites have fostered.

Today, when cities are striving to combat the harsh effects of climate change and health risks due to pollution, values of aesthetics posed as mere beautification can become a less-loaded dialogue. Etteger, Thompson, and Vincenzotti (2016) draws attention to how aesthetics is now less discussed when compared to functionality and sustainability of landscape architecture. The ideal urban green landscape today is a mix of amenity trees whose benefits range from merely serving as a visual delight, to those that are appreciated for being more purposeful of giving shade and cooling the environment; notwithstanding again, how not all benefits will be weighted equally. Maintaining this line of thought, Elizabeth Meyer argues in favour of making the links between aesthetics and sustainability in landscape architecture (Meyer, 2008 cited in Abbott, 2018). Combining both the utilitarian and visual experiences of landscapes, Abbott (2018) discusses how "productive landscapes" such as tourist destinations, farmlands or ecological habitats can add social, economic, and environmental elements of sustainability to landscape designs. Very much along these lines, the grammar of knowledge production around small-scale nature also benefits from valuing these landscapes in terms of their ability to bring out sensory perceptions and connect people to nature through intangible ways. To delve further on the aspect of productivity, Chapter 5 on Collectives which follows, will discuss some of these socio-cultural, economic, and ecological benefits which the landscape fractions offer in cities.

Nature that is non-purposive can appear beautiful too. Landscape fractions are often a combination of elements that are not purely architectural but pieces of it – created by a collaboration of actors such as industrial designers, mechanical engineers, carpenters and so on. Not just colonial heritage, but also industrial elements which have remained and have been transferred as a heritage to the post-industrial population, tend to "show up" amidst the wild growth, together creating an aesthetic and appeal that can be described as what Abbott (2018) words as "strangely familiar". This familiarity can come from memory values as highlighted by Etteger, Thompson, and Vicenzotti (2016): through the discourse of iron cast, train tracks of New York's Highline project that transformed a decommissioned overhead light rail transit line into a public park. The railway lines no longer serve any purpose but form part of the landscape design and contribute to what landscape critiques term as "experiences of the sublime" (Etteger, Thompson, & Vicenzotti, 2016).

FIGURE 4.8 Aesthetics of post-industrial sites: spontaneous nature blends with human-built objects inside soap factory industrial yard. January 14, 2020.

While the practice of landscape architecture contributes to the creation of new nature, the classification of landscape fractions adds to the vocabulary of the landscape architect, a set of relationships that bind aesthetic, and sometimes, non-aesthetic properties within a landscape which has been explained through the concept of "supervenience" by Zangwill (2007, p. 37) – it holds that certain aesthetic properties are dependent on non-aesthetic properties. In the final chapter of this book on Imaginaries, I will discuss how along similar lines, some architects have experimented with industrial elements to create a new definition of spaces. These experiments include repurposing and redesigning industrial spaces to match present-day requirements.

Notes

1 Cambridge University Press, 2023. *Landscape*. In: Cambridge Dictionary, available at: https://dictionary.cambridge.org/dictionary/english/landscape accessed on January 24, 2023.
2 Cambridge University Press, 2023. *Fraction*. In: Cambridge Dictionary, available at: https://dictionary.cambridge.org/dictionary/english/fraction accessed on January 24, 2023.
3 See Collier (2014).
4 Other landscape scholars have also pointed to how urban (wild) spaces are formed through informal changes which owners and managers make on the site, combined with natural processes (see Jorgensen & Lička, 2012).
5 H. Vangara. Effects of plant growth on buildings, interview with A. Deb, January 11, 2022.
6 Sir Banister F. Fletcher's works on History of Architecture has detailed illustrations of motifs and ornamentations compared across several cultures dating anywhere between 300 BC or early Egyptian architecture till the end of British colonial period (Fletcher, 1996). While floral elements have been used in architecture and design throughout history, a few iconic examples are: the Corinthian orders of the Greek (c. 900–27 BC) and the Roman (c. 509 BC to 400 AD) architectural periods; the use of floral engravements in furniture and decors during Baroque period spans between the early seventeenth to late eighteenth century spread across Italy, France and other countries in Europe; also originating in Europe, art nouveau refers to a shorter period between 1890 and 1910 which used botanical elements such as flowers and vines to accentuate spiral shapes in architectural design.
7 H. Vangara. Effects of plant growth on buildings, interview with A. Deb, January 11, 2022.
8 American naturalist Aldo Leopold pointed to the diversity in taste of people regarding aesthetics of nature (Leopold, 1966). Parsons (2008) further suggests Leopold's professional background as a scientist was not as essential per se, to perform the role of a critic, but rather his act of drawing from multiple "relevant scientific disciplines" (p. 52).

9 In introducing the concept of *stimmung*, Riegl (1903) separates the purely visual from moods and feelings of the observer. Arrhenius (2004) revisits Reigl's concept of stimmung as *visual directness*.
10 A. Shashwat. Photo-elicitation of landscape fractions with A. Deb, June 20, 2019.
11 N. Luke. Photo-elicitation of landscape fractions with A. Deb, June 21, 2019.
12 More on trajectory of abandoned spaces is discussed in Chapter 6 on Co-production.
13 This is in line with views of more experienced scholars in the discipline. The individual chapters of Catherine Dee and Katy Mugford in Jorgensen and Keenan's edited book delve more into these aspects (see Jorgensen & Keenan, 2011).
14 The High Line project in New York and Drosscapes in Germany are where people have deliberately superimposed vegetation to revive these areas. However, nature here was designed, while the built legacy was retained.

References

Abbott, M. (2018). The sustaining beauty of productive landscapes. *Journal of Landscape Architecture*, 13(2), 8–19. doi:10.1080/18626033.2018.1553389.

Arrhenius, T. (2004). The cult of age in mass-society: Alois Riegl's theory of conservation. *Journal of Historic Preservation, History, Theory, and Criticism*, 1(1), 75–81.

Bélanger, A. (2002). Urban space and collective memory: analysing the various dimensions of the production of memory. *Canadian Journal of Urban Research*, 11(1), 69–92.

Collier, M. (2014). Novel ecosystems and the emergence of cultural ecosystem services. *Ecosystem Services*, 9, 166–169. doi:10.1016/j.ecoser.2014.06.002.

Doherty, G. (2017). *Paradoxes of Green: Landscapes of a City-State*. Oakland, CA: University of California Press.

Dunin-Woyseth, H. (1996). Reviewed work: The city of collective memory: its historical imagery and architectural entertainment by M. Christine Boyer. *The Town Planning Review*, 67(1), 124–126.

Edensor, T. (2005). *Industrial Ruins: Space, Aesthetics and Materiality* (1st ed.). Oxford: Berg.

Etteger, R. v., Thompson, I. H., & Vicenzotti, V. (2016). Aesthetic creation theory and landscape architecture. *Journal of Landscape Architecture*, 11(1), 80–91. doi:10.1080/18626033.2016.1144688.

Fletcher, B. F. (1996). *A History of Architecture on the Comparative Method for Students, Craftsmen & Amateur* (5th ed.). London: Architectural Press.

Gandy, M. (2013). Marginalia: aesthetics, ecology, and urban wastelands. *Annals of the Association of American Geographers*, 103(6), 1301–1316.

Gandy, M. (2016). Unintentional landscapes. *Landscape Research*, 41(4), 433–440. doi:10.1080/01426397.2016.1156069.

Jorgensen, A., & Keenan, R. (Eds.). (2011). *Urban Wildscapes* (1st ed.). New York: Routledge.

Jorgensen, A., & Lička, L. (2012). Anti-planning, anti-design? Exploring alternative ways of making future urban landscapes. In A. Jorgensen, R. Keenan, A. Jorgensen, & R. Keenan (Eds.), *Urban Wildscapes* (pp. 221–236). New York: Routledge.

Kaplan, R. (1975). Some methods and strategies in the prediction of preference. In R. O. E. H. Zube (Ed.), *Landscape Assessment: Values, Perceptions, and Resources* (pp. 118–129). Stroudsburg, PA: Dowden, Hutchinson, & Ross.

Knez, I., Sang, Å. O., Gunnarsson, B., & Hedblom, M. (2018). Wellbeing in urban greenery: the role of naturalness and place identity. *Front. Psychol.*, 9, 1–10. doi:https://doi.org/10.3389/fpsyg.2018.00491.

Krapez, A., Hughes, M., & Newsome, D. (2021). Perceptions of 'naturalness' by urban park visitors: insights from Perth, Western Australia. *International Journal of Geoheritage and Parks*, 9(4), 450–462. doi:https://doi.org/10.1016/j.ijgeop.2021.11.004.

Lak, A., & Hakimian, P. (2019). Collective memory and urban regeneration in urban spaces: reproducing memories in Baharestan Square, city of Tehran, Iran. *City, Culture and Society*, 18(1), 100290.

Leopold, A. (1966). *A Sand County Almanac, with Essays on Conservation from Round River*. New York: Ballantine Books.

Lo, A. Y., & Jim, C. Y. (2015). Community attachment and resident attitude toward old masonry walls and associated trees in urban Hong Kong. *Cities*, 42 (A), 130–141. doi:10.1016/j.cities.2014.09.006.

Mandelbrot, B. B. (1982). *The Fracral Geometry of Nature* (Rev. ed. of: *Fractals* c.1977 ed.). New York: W. H. Freeman and Company.

Meyer, E. (2008). Sustaining beauty. The performance of appearance: a manifesto in three parts. *Journal of Landscape Architecture*, 3(1), 6–23. doi:https://doi.org/10.1080/18626033.2008.9723392.

Mill, J. S. (1874). Nature. In *Nature, the Utility of Religion, and Theism* (2nd ed., pp. 3–68). London: Longmans, Green, Reader, and Dyer.

Parsons, G. (2008). *Aesthetics and Nature*. London: Continuum International Publishing Group.

Riegl, A. (1903). *Der Moderne Denkmalkultus: Sein Wesen und seine Entstehung* (English translation by Kurt W. Forster and Diane Ghirardo published in 1982 by Oppositions 25 ed.). Vienna: Braumüller.

Robles, K. E., Roberts, M., Viengkham, C., Smith, J. H., Rowland, C., Moslehi, S., … Sereno, M. E. (2021). Aesthetics and psychological effects of fractal based design. *Front Psychol.*, 12(699962). doi:10.3389/fpsyg.2021.699962.

Rose, G. (2001). 'The good eye': looking at pictures using compositional interpretation. In *Visual Methodologies: An Introduction to Interpretation of Visual Materials* (pp. 33–53). London: SAGE Publications.

Ryan, J. C. (2013). Botanical memory: exploring emotional recollections of native flora in the Southwest of Western Australia. *Emotion, Space and Society*, 27–38. doi:10.1016/j.emospa.2012.09.001.

Spehar, B., Clifford, C. W., Newell, B. R., & Taylor, R. P. (2003). Universal aesthetic of fractals. *Computers & Graphics*, 27(5), 813–820. doi:https://doi.org/10.1016/S0097-8493(03)154–157.

Taylor, R. (2006). Reduction of physiological stress using fractal art and architecture. *Leonardo*, 39(3), 245–251. doi:https://doi.org/10.1162/leon.2006.39.3.245.

Vaughn, A. C. (2015). *Aesthetics and Performance Evaluation of PostIndustrial Public Parks.* Landscape Architecture Undergraduate Honors Thesis, University of Arkansas, Fayetteville.

Zangwill, N. (2007). *Aesthetics Creation.* New York: Oxford University Press.

5

COLLECTIVES

Invisibility of everyday nature

Consider this scene: a man carrying a sack of potatoes walks through the lanes of Kumartuli, located only minutes away from the banks of Hooghly River, a tributary of the river Ganges, along which the city of Kolkata is built. He hardly notices the natural bouquet of plants growing on the walls. He walks this path every day and perhaps knows these plants exist even when he is looking elsewhere. He would have probably seen the plants thrust against the masonry, any other day. But today, he can't wait to reach home (or his restaurant) and unload the heavy sack of potatoes. Perhaps he doesn't have time amid the busy day to appreciate the beautiful bouquet of plants grown against the mercantile red brick masonry. But perhaps, he is also not able to appreciate, because the connections which this assemblage has established with the entire city, is invisible to the eyes.

Kumartuli is one of the oldest neighbourhoods of Kolkata. Some houses in Kumartuli date before the British colonial period (i.e., c. 1800s), built for merchants who needed large volumes of space to store goods that were shipped through the river Ganges. The walls thus reek of masonry prevalent during the time. But in more recent decades, parts of the neighbourhood have been subject to outmigration and physical decay. Moreover, it has become extremely hard for residents to maintain

DOI: 10.4324/9781003374602-5

the massive structures that depended on revenue from the mercantile economy for their upkeep. Many buildings have worn-out plaster, exposing the intricate masonry of arches and bricks. The economy of the area has been reduced from a bustling logistical point to a handful of traditional workshops and food outlets. Given the aged built fabric, one should not be surprised to spot a few wild plants and even full-grown trees growing on these buildings.

Kumartuli's sculpture of wild growth on brick masonry stands alone, but within its class of hundreds of thousands, if not millions, of such small-scale arrangements spread across the town (Figure 5.1). Although minimal in its appearance, if one begins to map benefits of biodiversity, climate, livelihoods, and even aesthetics – it is likely to be found that even these tiny fragmented assemblages which I refer to as "landscape fractions" perform as collective ensembles, interconnected with each other, to offer the aforementioned ecological services. Perhaps one reason that makes it difficult to picture these small-scale wild spaces performing as a whole is because they are not physically connected by continuous

FIGURE 5.1 A bouquet of plant growth on building walls of Kumartuli in Kolkata: the plaster has fallen off to expose the red brick masonry of the mercantile period. July 6, 2018.

threads of green, but through temporalities such as grazing patterns of livestock and foraging paths of bird species; or even invisible media such as atmosphere and eyesight.

Some benefits of small-scale nature may not be apparent when fractions are viewed individually. But ecosystem services of multiple fractions can combine to deliver unique benefits which individually, assemblages may fall short of. The collective actions of small-scale nature that lies fragmented across urban landscapes are difficult to perceive without engaging in an imagination across biological and socio-economic processes within the city. Some services that may also be insignificant at the site level might become significant when multiplied across the city. Other times, fractions within a locality might form an inter-connected network supporting non-human life or compose a visual assemblage that enhances the appeal of a place. It is perhaps relevant, then, to assess the values of these landscapes at different scales. This chapter will discuss four classes of benefits which cities receive from small-scale wild spaces.

Food, flowers, and fodder for livelihoods

On an early winter morning, when the sun was yet to rise fully upon Bengaluru, a herd of ten cows moved through the quiet streets of a residential neighbourhood. Typical mornings here are cold, and the herder had probably decided to stay indoors. In charge of navigating themselves, the herbivores walked lazily to begin their journey for the day; and the journey is set to follow the intricate network of neighbourhood streets. Plants growing along road verges or that have grown over the fences of properties are potential grazing sites. The streets are connected to larger pockets of green such as wild grass growing on vacant properties and parks. Together these compose the conduits of flow or the "grazing network" that urban herders bank upon for feeding their cattle. Cattle rearing in Indian cities is seen both in the outskirts as well as in the central areas of the city. Sometimes, households rear a sheep or two for subsistence, and other times herds of 10 to 20 sheep or goats as part of full-scale commercial production of meat. Unlike goats and sheep, cows and buffaloes are typically reared for milk and not meat, given their widely held religious connotations. Thus, broadly speaking, cattle rearing in cities reduce food miles and offer livelihoods with an easy skill level and a low investment amount for people who are in the business.

Agrarian activities are perceived to belong in rural areas rather than in cities. Town planning regimes in India have moved agrarian activities as well as industries to the outskirts of cities. However, small pockets of agricultural produce continue to thrive, often informally within cities. Tourists from outside of India are often amazed at how cows roam the busy streets in Bengaluru. Academics internationally have been curious to know how goats in Kolkata thrive on food waste generated from markets and households – in turn serving as much needed scavengers in the waste management system. In reality, livestock rearing is not at all rare in Indian cities, offering a substantial source of income. Small-scale dairies and family-run butcher shops are not uncommon. Moreover, city administrative boundaries have been expanded into large tracts of rural land, where lifestyles are yet to fully transform from rural to urban. Thus, cattle rearing remains an important activity, particularly in the peripheries of Indian cities.

Livestock is not only reared for commercial purposes, but for some it is a mode of subsistence; it offers a safety net for otherwise dependent individuals. We return to our park in Bengaluru where the children had lost their cricket ball: situated on the outskirts of a rapidly expanding metropolis, this public asset serves as a grazing ground for local herders. The giant trees with spread out foliage in the park offer shade for sheep and the shepherd to rest after traversing long routes throughout the day, making it quite an oasis in middle of an otherwise concrete neighbourhood. One can simply walk into this oasis without opposition as the gates remain open without any monitoring from public authorities; and local residents who use the park do not seem to mind either. An elderly woman would come every day into this neighbourhood park with her two sheep. Her wrinkled skin and grey hair, with a slight slouch of the back were perhaps give-away signs that she was well above her working age. She would lazily walk around the park where tall patches of grass had grown amid the otherwise impoverished land, bared down to its soil due to neglect from the maintenance authority. Her two mature sheep calmly grazed on these patches or fractions of plant growth and refrained from misbehaving by running hither and thither. The old woman and her two sheep lived in a low-income community that have set up their make-shift shelters nearby.

Following typical cultural trends in India, women may have been homemakers all their working lives with little savings at their disposal when they are old. Thus, many at times may be largely, if not entirely, dependent on care from younger family members, disrupting the image of a more recent social change towards nuclear families in cities. At the same time, the responsibility of such pastoral activities often rests with members of the family who are not otherwise employed as breadwinners – in many instances they are either children or women. Writing on time-use and activities in Bengaluru Metropolitan Region, Deb and Kidwai (2020) have explained this phenomenon as "time-compression", where labour is socially distributed within family members for increasing household productivity. Each sheep might fetch her INR 30,000 (USD 411 at current prices)[1] from the market, which enables her to contribute a significant annual income to her family. The income from rearing livestock may contribute to everyday expenditures of her own or even act as a fund for medical emergencies. Then the network of street-side nature (Figure 5.2), vacant land and neglected parks all serve as a support system for grazing – enabling an alternative income, particularly for the low-skilled population.

FIGURE 5.2 Streetside nature serving as grazing patch in Bengaluru.

Data on occupation is hard to get because of the pace of urbanisation – but only a handful of the houses in this neighbourhood actually own cattle that stray (confidently, less than one per cent). Other Indian towns such as Ahmedabad may have higher numbers. Increases in the numbers of cattle can, in fact, threaten the plant population in a neighbourhood. For instance, when not accompanied by herders, cattle may stray into private yards and damage designed landscapes (Shackleton, et al., 2017). They may also destroy newly planted tree-saplings on the street side. Overgrazing of street-side grasses and plants exposes soil and leads to soil erosion. Furthermore, the dung from cows and goats are considered to be problems of urban blight as well as of attracting insects that cause irritation to residents. However, experts argue that this may be more a management problem rather than a problem with the idea of grazing itself.[2] For instance, cities have experimented with biogas[3] plants to manage animal excreta and protecting plants by the use of caging or fencing trees, planting thorny and less palatable plants to reduce damage to trees.

Perhaps a worthwhile mention to this discussion, would be foraging of flowers and vegetables. Wild plants growing in the backyard or street sides can have edible growth to offer to those who are able to find it. Mushrooms and berries in the UK and pumpkins, herbs, tomatoes in India are only a few products I came across. They depend on perceptions of health: when I found a fragrant patch of Fenugreek herb (*Trigonella foenum-graecum*), growing along the street side of a quaint residential neighbourhood, I couldn't help but enquire if people in the nearby households foraged them for cooking. One woman pointed out, "Back in our village, we picked them from fields for cooking. But these have grown near the drain, and the soil is polluted. We are not comfortable consuming them and rather buy them from the shops".[4] On the other hand, flowers and decorative plants are less risky to be used and foraging for them is rather commonplace in Indian cities.

The intensities in which urban dwellers go foraging for flowers and vegetables can be categorised into several gradients: on the deeply incidental end, children may be interested to bring home their discoveries; a slightly more methodical approach is when homemakers and cooks may keep a lookout around their own homes, for fresh grown vegetables and herbs, to add to their cooking or when nurserymen spot and collect decorative plants[5] to put them on sale; and a more organised activity would be hobbyists forming groups that share information and plan excursions to hunt for produce. There are now mobile phone applications which facilitate organised information sharing for foraging.

Wellbeing without eco-gentrification

Greening through amenities such as parks and avenues can sometimes accelerate gentrification within a neighbourhood, especially in cities like Bengaluru that have noticeable disparities. Bengaluru in recent decades has experienced economic restructuring that has also led to a wide income gap between different groups. The city that was earlier thought to be a "Garden City" and a "Pensioner's Paradise" is now known as the "Silicon Valley of India" (Sastry, 2013). With the information technology (IT) sector and knowledge economy growing in the city, there is a stark disparity of income between the skilled workers and others. This has directly affected the shaping of the city's fabric. Developers are now building new neighbourhoods with more amenities that offer extravagant lifestyles for those who can afford them – and fetch a higher profit. In such neighbourhoods, it has been typical to find informal bazaars being replaced by malls and supermarkets that manifoldly increases the cost of living. While urban green infrastructures such as public parks and street-side landscaping (as well as private amenities like backyards and rooftop gardens) are good for people's wellbeing, they also raise real estate values and demand costs of maintenance.[6] Unable to afford the rising rents and living costs, less affluent residents have often had to move out of such neighbourhoods to set up homes in less developed areas.

However, a suburban neighbourhood, very close to Bengaluru's industrial district, Peenya, is withstanding this pressure, while still ensuring a quality life for its residents. While the residential layout has been developed with little space for trees, patches of green have emerged between buildings. These patches offer bursts of green along the road as opposed to one dominant central green space (Figure 5.3). The informal or unintentional growth of plants (trees, shrubs, and grasses) in residual spaces mixes with other, more ornamental, plants grown intentionally by people in their own properties to form a greener streetscape. In this mix of formal–informal assemblage, both public and private actors have facilitated the landscape: while the municipality has treelined the street sides with jacaranda (*Jacaranda mimosifolia*), Caribbean trumpet tree (*Tabebuia aurea*), and pink trumpet (*Tabebuia avellanedae*), households along the street have also planted trees in front of their homes independently – be it a Cook pine (*Araucaria columnaris*) or the jhau tree (*Tamarix dioica*). And along the same street, there are also plants which have grown on their own. However, when one looks at the assemblage, it is hard to tell which plants are wild and which are not. One might as well ignore assessment of the individual contributions of human and non-human forces here, and instead appreciate the beauty of the collective landscape.

Scholars such as Del Tredici (2010) and Gandy (2013) have brought attention to how site and contextual setting create meanings of urban nature. To reiterate a few of their points discussed in earlier Chapters 1, 2, and 3: vegetation growing in a degenerating neighbourhood becomes perceived as a sign of decay (Del Tredici, 2010). Moreover, urban nature in sites of decay often grows on abandoned buildings, sometimes damaging them and hence nature becomes counter-utilitarian (Gandy, 2016). Thus, it is the larger landscape – which in the case of Figure 5.3 is a thriving neighbourhood – that has created a positive context for the collective of landscape fractions. I will discuss more on how meanings of small-scale spontaneous nature can be shaped through context in Chapter 7 on Frames. But it is perhaps worth noting now, that despite possible negative connotations associated with wild spaces in cities, research over the last 30 years has broadly found nature to have a positive effect on people's mental health and wellbeing (Loder, 2020; Carrus, Lafortezza, Giuseppe Colangelo, Scopelliti, & Sanesi, 2013). Two notable literature examples, which those in the discipline might be familiar with, are that of Kaplan and Kaplan (1990) as well as Ulrich, et al. (1991) which highlight values of stress reduction and psychological restoration of human beings in natural environments. Subsequently, others have extended this connection between human health and nature, to test more specifically, if wild character or "naturalness" impacts the results: in the city of Bari (Italy), Carrus, et al. (2013) found that visuals of urban landscapes with more naturalness positively influenced people's restoration. In a more controlled set-up and using audio-visual methods, Berg, Jorgensen, and Wilson (2014) also found a strong significance of stress reduction and naturalness. However, it is also worth noting that the studies used photographs of large-scale settings such as botanical gardens, urban parks, and woodlands that are representative of green spaces both within the city and its peri-urban areas.

Now let us return to our streetscape where the formal and informal green spaces along the street appeared as a collective assemblage to a

FIGURE 5.3 Wellbeing without Gentrification: spontaneous nature complements formal landscaping to make a neighbourhood in Bengaluru, greener. This benefits health and wellbeing of residents with no additional cost.

passer-by walking down the street. It would be apparent to any observer how the wild grasses and plants growing on the unmanaged verges, and at times overflowing on to the street, have added to the liveability of the neighbourhood. In this example of a neighbourhood street, nature existed as an ensemble with simple human-made elements of the road, the drain cover, and pavements. The residents have incurred very little costs to maintain the landscape and yet enjoy its many benefits. The low cost of living in the neighbourhood has allowed industrial workers to own a home and live near their workplace. Yet such health benefits are not always prioritised, and urban designs are being implemented in cities that facilitate a new aesthetic, wiping out "messy" small-scale nature or ground cover through "white topping" of streets and plazas. Before and after comparison in Figure 5.3 shows how a neighbourhood streetscape loses its visual tranquil after being stripped of spontaneous growth through acts of de-weeding and repairs.

Our understanding of connectedness of people with nature has come a long way since E. O. Wilson's *Biophilia* hypothesis in 1984 that claimed that humans have an innate desire to be drawn to nature (Wilson, 1984). More recent research has highlighted the crucial role of local or nearby nature in ensuring human wellbeing (Kaplan, 1992); suggesting that for healthy communities it is important to have nature as close as possible to everyday experiences of urban spaces. Yet, the consequences of eco-gentrification and scarcity of resources (e.g., land, water, labour) discourage planners and architects from allocating green spaces in low-income neighbourhoods. However, taking cues of the *Just Green Enough* approach to maintaining green spaces in the city, scholars have suggested informal growth of nature as a mode of greening (Rupprecht & Byrne, 2017). Discussions with practitioners reveal that contrary to designed nature, informal nature has little or no impact on rising real estate values,[7] thus implying a low risk of eco-gentrification.

Micro-habitats and stepping stones

The experience of naturalness within a neighbourhood is heightened by the songs of birds, croaking of frogs, and rubbing of crickets; and more trees (and plants) are generally good for birds, insects, and other animals. Small-scale nature is home to a host of biodiversity – both in terms of flora and fauna. Minimal nature which can range from a tree on a rooftop to wild grass growing out of cracks in a wall that can become home to insects, birds, and small animals. One tree growing next to a partly abandoned mansion in Kolkata has several nests of birds such as myna and barbet. Citizens observe these birds have "returned to the city" or have become more visible, after the city government decided to increase Kolkata's tree cover. In Bengaluru, the open drains have offered a habitat for pond herons (Figure 5.4). The tiny fishes and tadpoles in these waters become food for birds. In certain neighbourhoods, recent efforts of the local government to clean up some of these drains have resulted in newly dredged soil that have temporarily exposed earthworms and other sub-soil fauna – also food for birds. Noticeably, herons here have also increased in numbers.

Upon looking closer, small patches of grasses or plants have a wide diversity of floral species. Small-scale spontaneous growth are like islands of nature, serving as micro-habitats that are permanent homes of animal life and, at other times, serving as stepping stones between two large green pockets (Vega & Küffer, 2021). Therefore, in landscape

FIGURE 5.4 A heron searches for food in a dredged storm water drain in Bengaluru. Herbaceous plants have grown along the drain to add green to the site, imitating a pond bank.

fractions, bugs, caterpillars, and many other insects establish their own territories (see Figure 5.4 and Figure 5.5). Small insects can make landscapes interesting to observe – think of the colourful ladybugs and butterflies or the energetic grasshoppers in the wildflower bushes. Or, for instance, the algae-infested open-to-air water treatment tanks of the soap factory where one can find bees, dragonflies, and water striders on the water surface as well as guppies and tadpoles beneath it. The city is dotted with such natural pockets that – upon zooming out at the city scale – create a mosaic of diverse novel habitats.

Returning to one of the more under-represented small-scale settings of spontaneous nature, and also one of the focuses of this book – building or compound walls can accommodate a wide variety of habitats. Microplants such as fungi, algae, cyanobacteria, moss, slime mould, and bacteria grow on walls. Even a thin layer of growth such as moss beds or fungus mats may provide anchorage needed to create conditions of refuge for invertebrates and insects, which may further attract birds.[8] Then there is also ivy, fern, and grasses or even trees, that grow on masonry to provide a range of habitats across the city. The structure and its masonry together form an interesting terrain which may offer sunbasking platforms, roosting sites, or refuge for rare species of organisms (Cathersides & Parker). An excerpt from a report by English Heritage helps understand the seasonal dynamism of such habitats created by both built and natural elements of the assemblages:

> *Wall vegetation, especially Ivy, provides a habitat for animals to live in and serves as places for hibernating butterflies and moths in winter, and sites for birds to nest during the spring. Ivy provides further benefit as it flowers and fruits late in the autumn and winter when other flowers and berries are not available as food for insects and birds.*
>
> *(Cathersides & Parker, p. 4)*

Wall settings are far from boring and not enough work has been done to identify their values; at least in India. However, longevity of such biological growth depends on water retention capacity of the walls and are vulnerable to drying out during summers.

Moreover, plants considered weeds, may also attract birds and pollinators. For instance, *lantana camara* is well known to be an invasive species, and cities are struggling to control its spread. However, the

FIGURE 5.5 Micro-habitat (a): Varying densities (red dots) of red cotton bugs along a narrow strip along the roadside in Bengaluru.

FIGURE 5.6 Micro-habitat (b): The narrow strip with red cotton bugs, in Bengaluru.

flowers of this plant attract butterflies and other pollinators. But it must be noted that micro-habitats can harbour insects and other micro-organisms which can pose risks to human health (such as mosquitoes). Diseases such as malaria, dengue fever, and chikunguniya spread through mosquitoes which find their habitats in stagnating water in stray tyres, puddles, neglected water storage vessels, or overgrown plants which Rehman (2018) captures as "micro-ecologies" urging that such sites require greater surveillance for public health, especially in the poorer neighbourhoods (see also Gandy, 2014; Swyngedouw, 2006). In India, Malaria alone caused about 7,700 deaths (2019 estimate) in a year; but those numbers have drastically reduced by about three-quarters since the 2000s (PTI, 2020). Yet the severity of negative consequences perhaps overshadows some of the cultural and ecological benefits of these sites.

On the other hand, an emerging body of experiments in immunology is looking at how certain micro-organisms in the urban biome can be beneficial for the human body (Mills, et al., 2017). Jamie Lorimer hypothesises a "probiotic" turn in biology (opposite of antibiotic) where interaction with microbes would be encouraged to thrive on the human body – for the benefits of developing immunity, etc. (Lorimer, 2020). If the experiments generate enough evidence, then it would open up possibilities of designing small-scale nature in cities with microbes that are beneficial to human health.

Small-scale nature or fractions of botanical growths can serve as habitats for insects – and this is no small a contribution. The insects, along with fruits from wild plants (Figure 5.7), further become food for birds – thereby supporting more visible and appreciated wildlife within the city. At first instance, landscape fractions might seem to provide island habitats for lifeforms. But upon viewing such dotted growth as parts of the wider urban green lattice, they can be seen to connect "larger islands" by becoming stepping stones in between. Take for example, the isolated tree that has wrapped around a building as if unwilling to let go of its last strand of hope, but then becomes resting grounds for birds that are hopping from one urban garden to another. The grasses that have grown from cracks of the pedestrian walkway, becoming umbrellas for trails of ants and bugs underneath that use it to travel between habitats separated within the city. And it is not uncommon for snails to otherwise roast under the sun without being able to find a patch of shaded bush to rejuvenate under.

FIGURE 5.7 Wild climbers with berries growing on park fence in Bengaluru. Berries, and other fruits, if edible, may become food for birds and animals.

Regulating the urban environment

Cities are facing risks of increasing air pollution due to emissions. Heat emitted by reflective and paved surfaces within cities also increase temperatures and exacerbates effects of heatwaves. Rapid climate change has, in many cases, challenged cities to adapt to increased rainfall. The loss of green cover compounds these problems and particularly puts greater pressure on city's hard infrastructure such as storm water drains. Certain neighbourhoods get affected disproportionately, when they do not have the required resources to plan more green spaces (see scarcity of resources in Chapter 2 on Meanings). Wolch, Byrne and Newell (2014) argue the approach of Just Green Enough, whereby neighbourhoods can utilize their limited resources to plan for small-scale green spaces – they may not offer a full range of benefits, but nevertheless provide with some benefits rather than nothing. Architects may do this through roof gardens and urban designers may create pocket parks. There cannot be a one size fits all approach as certain neighbourhoods may not have space, others may have space but not enough resources (saplings, water, labour, etc.) for maintenance. It is also in these oddly shaped spaces that are insufficient to be appropriated for standard designs, where fractions of spontaneous nature emerge.

At first, landscape fractions appear too small to offer substantial climate regulation at the local level. Indeed, a small patch of grass will not be enough to stop flooding of the area, despite its property of reducing runoff. They may not cool a neighbourhood enough[9] to offer climate comfort despite properties of evapotranspiration. However, say at the city level, small-scale nature may add-up: in Greater London for instance, 7.5 per cent of ground cover is classified as wild grass (Rogers, Sacre, Goodenough, & Doick, 2015). Incidentally, one study conducted in Manchester, UK, also found that in the urban setting, grass is more effective in absorbing rainfall and runoff than trees (Armson, Stringer, & Ennos, 2013). Wild urban spaces would offer the same climate regulatory functions like carbon removal for air purification and cooling for mitigating the effects of urban heat islands (UHIs). Perhaps it is not prudent to dismiss roles of landscape fractions just because of their small-scale stature. Notably, it is found that in terms of relative contribution between the designed and wild landscapes, patches of spontaneous nature bearing a greater diversity and lesser chemical influence on their natural functions, offer better climate regulation per unit area. Vieira, et al. (2018) write:

> *vegetation type characterized by a more complex structure (trees, shrubs and herbaceous layers) and by the absence of management (pruning, irrigation and fertilization) had a higher capacity to provide the ecosystems services of air purification and climate regulation. By contrast, lawns, which have a less complex structure and are highly managed, were associated to a lower capacity to provide these services.* (p. 306)

Thus, it follows at least in theory, every wild patch lost and replaced with designed landscape makes the urban forest a little less efficient in terms of regulatory values. Having said that, effectiveness of collective action would depend on the threshold of services required within each regulatory type: for instance, small-scale spontaneous growth may help reduce runoff and alleviate pressure on drainage infrastructure, but may not be enough to reduce flood risk in the event of heavy rain or in the absence of adequate drainage systems.

Sum greater than parts

The value of landscape fractions is more than what immediately meets the eye. Despite being sites of dereliction and decay, landscape fractions support livelihoods, biodiversity, and climate regulation along with offering cultural benefits. However, the fragmented distribution of small-scale spontaneous nature may mean that these benefits are not offered significantly at the site level. Instead, they may function collectively to deliver ecosystem services at the neighbourhood or perhaps at the urban scale. The understanding of how these natures offer collective services is a complex web of interaction. Not every landscape will offer the same set of benefits – for instance, while all sites broadly contribute to climate regulation in differing capacities, not all sites may offer suitable animal habitats or fodder. In addition, not all landscape fractions may combine to make a landscape look "whole". This makes an interesting constellation of benefits – that goes beyond what is visible – urging urbanists and ecologists to identify various types of networks which assemblages are part of; these may include grazing networks and networks of stepping stones.

Therefore, the study of landscape fractions calls for an evaluation of small-scale wild spaces as collective under the broad class of urban forest, rather than valuing each site in isolation. When the sites of interest are looked at individually, their importance is reduced; mainly because such does not offer a complete understanding. But when viewed as collective ensembles (e.g., networks) in conjunction with space, time, and atmosphere, a new set of roles are revealed and the small-scale botanical growths can be accounted for offering a wide range of ecosystem services. These services are but among several crucial roles of landscape fractions in making cities habitable for both human and non-human lifeforms.

Amid a struggle to save urban nature from being bulldozed into apartments and malls with little green cover, these sites have contributed to keeping cities healthy, aesthetic, and vibrant. Yet in most instances, they have faced tough maintenance regimes that look to combat spontaneous nature through weed management or soil sealing, leaving little room for natural growth. If the scenario continues as usual, the future looks grim; and perhaps worth keeping the broad picture in mind, that suggests, in the coming two decades urban air pollution and UHI effect are, respectively, estimated to cause 6.6 million and 250,000 premature deaths annually across the world (see Lelieveld, Evans, Fnais, Giannadaki, & Pozzer, 2015 and WHO, 2014 cited in Vieira, et al., 2018). Thus, the cost of losing nature within cities is high, within which landscape fractions play a role. Yet, no stock of spontaneous vegetation is taken. Considering that replacement costs of any kind of nature is high – as municipalities in several cities are unable to deploy resources like water due to scarcity, or labour due to shortage of funds – it is only prudent to consider values of nature that has thrived outside of human control and are shaping built-environments around them.

Notes

1 As per regional estimates, each sheep may fetch anything between INR 17,000 to INR 25,000 (Shourabh, 2019).
2 B. Banerjee. Wild growth and urban planning, interview with A. Deb, December 9, 2021.
3 B. Banerjee. Wild growth and urban planning, interview with A. Deb, December 9, 2021.
4 Informal field discussion with a homemaker, 2019, Bengaluru, India.
5 Informal field discussion with nurserymen, 2022, Bengaluru, India.
6 One study from New York's High Line project suggests a 35.3 per cent increase in real estate value (Black, 2020). The paper also cites useful literature that points to gentrification and displacement of the existing residential population.
7 B. Banerjee. Wild growth and urban planning, interview with A. Deb, December 9, 2021; E. Longchar. Attitudes towards botanical growth, interview with A. Deb, January 2, 2022.
8 J. Munirathinam. Ecology of small-scale wild growths in cities, interview with A. Deb, November 23, 2021.
9 In understanding the potential of small green spaces to regulate the urban microclimate, a study by Park, Kim, Sohn and Lee (2021) indicate that small-scale nature can play a supporting role, alongside large green spaces; however, they are less effective in the absence of large green spaces.

References

Armson, D., Stringer, P., & Ennos, A. R. (2013). The effect of street trees and amenity grass on urban surface water runoff in Manchester, UK. *Urban Forestry & Urban Greening*, 12(3), 282–286. doi:https://doi.org/10.1016/j.ufug.2013.04.001.

Berg, A. E., Jorgensen, A., & Wilson, E. R. (2014). Evaluating restoration in urban green spaces: does setting type make a difference? *Landscape and Urban Planning*, 127, 173–181. doi:https://doi.org/10.1016/j.landurbplan.2014.04.012.

Black, K. J., & Richards, M. (2020). Eco-gentrification and who benefits from urban green amenities: NYC's High Line. *Landscape and Urban Planning*, 204, 103900. doi:https://doi.org/10.1016/j.landurbplan.2020.103900.

Carrus, G., Lafortezza, R., Giuseppe Colangelo, I. D., Scopelliti, M., & Sanesi, G. (2013). Relations between naturalness and perceived restorativeness of different urban green spaces. *PsyEcology*, 4(3), 227–244. doi:http://dx.doi.org/10.1174/217119713807749869.

Cathersides, A., & Parker, E. (2014). *Landscape Advice Note: Vegetation on Walls*. English Heritage. Retrieved January 3, 2023, from https://www.english-heritage.org.uk/siteassets/home/learn/conservation/gardens-and-landscapes/lan_-_vegetation_on_walls.pdf.

Deb, A., & Kidwai, A. H. (2020). Time-pressure: a planning imperative in the development process of a fast urbanising metropolitan region. In V. Yadav (Ed.), *Sustainable Regional Development*. New Delhi: Studera Press.

Del Tredici, P. (2010). Spontaneous urban vegetation: reflections of change in a globalized world. *Nature and Culture*, 5(3), 299–315.

Gandy, M. (2013). Marginalia: aesthetics, ecology, and urban wastelands. *Annals of the Association of American Geographers*, 103(6), 1301–1316.

Gandy, M. (2014). Mosquitoes, modernity, and postcolonial Lagos. In *The Fabric of Space: Water, Modernity, and the Urban Imagination* (pp. 81–108). Cambridge, MA: MIT Press.

Gandy, M. (2016). Unintentional landscapes. *Landscape Research*, 41(4), 433–440. doi:10.1080/01426397.2016.1156069.

Kaplan, R. (1992). The psychological benefits of nearby nature. In D. Relf (Ed.), *The Role of Horticulture in Human Well-being and Social Development* (pp. 125–133). Portland, OR: Timber Press. Retrieved from https://deepblue.lib.umich.edu/handle/2027.42/148476.

Kaplan, R., & Kaplan, S. (1990). Restorative experience: the healing power of nearby. In M. Francis, & R. T. Hester Jr. (Eds.), *The Meaning of Gardens* (pp. 238–243). Cambridge, MA: MIT Press.

Lelieveld, J., Evans, J. S., Fnais, M., Giannadaki, D., & Pozzer, A. (2015). The contribution of outdoor air pollution sources to premature mortality on a global scale. *Nature*, 525(7569), 367–371. doi:https://doi.org/10.1038/nature15371.

Loder, A. (2020). *Small-scale Urban Greening: Creating Places of Health, Creativity and Ecological Sustainability*. New York: Routledge.

Lorimer, J. (2020). *The Probiotic Planet: Using Life to Manage Life*. Minneapolis, MN: University of Minnesota Press.

Mills, J. G., Weinstein, P., Gellie, N. J., Weyrich, L. S., Lowe, A. J., & Breed, M. F. (2017). Urban habitat restoration provides a human health benefit through microbiome rewilding: The Microbiome Rewilding Hypothesis. *Restoration Ecology*, 25(6), 866–872. doi:10.1111/rec.12610.

Park, J., Kim, J.-H., Sohn, W., & Lee, D.-K. (2021). Urban cooling factors: do small greenspaces outperform building shade in mitigating urban heat island intensity? *Urban Forestry & Urban Greening*, 64, 127256. doi:https://doi.org/10.1016/j.ufug.2021.127256.

PTI. (2020, December 1). *India records largest reductions in malaria cases in South-East Asia, says WHO*. Retrieved January 4, 2023, from *The Hindu*: https://www.thehindu.com/sci-tech/health/india-records-largest-reductions-in-malaria-cases-in-south-east-asia-says-who/article33219416.ece.

Rehman, N. (2018). Mosquitoes and the urban political ecology of stagnant water. *AAG Annual Meeting 2018*. American Association of Geographers.

Rogers, K., Sacre, K., Goodenough, J., & Doick, K. (2015). *Valuing London's Urban Forest: Results of the London iTree Eco Project*. London: Treeconomics.

Rupprecht, C. D., & Byrne, J. A. (2017). Informal urban green space as anti-gentrification strategy? In W. Curran, & T. Hamilton (Eds.), *Just Green Enough: Urban Development and Environmental Gentrification*. London and New York: Routledge.

Sastry, N. N. (2013). Bangalore: from garden city to Silicon Valley of India. In R. P. Misra (Ed.), *Urbanisation in South Asia Focus on Mega Cities* (pp. 227–251). New Delhi: Cambridge University Press India Pvt. Ltd. doi:https://doi.org/10.1017/9789382993087.009.

Shackleton, C. M., Guild, J., Bromham, B., Impey, S., Jarrett, M., Ngubane, S., & Steijl, K. (2017). How compatible are urban livestock and urban green spaces and trees? An assessment in a medium-sized South African town. *International Journal of Urban Sustainable Development*, 9(3), 243–252. doi:10.1080/19463138.2017.1314968.

Shourabh, V. (2019, August 1). *Welcome To Sheep City*. Retrieved May 13, 2022, from Star of Mysore: https://starofmysore.com/welcome-to-sheep-city/.

Swyngedouw, E. (2006). Circulations and metabolisms: (hybrid) natures and (cyborg) cities. *Science as Culture*, 15(2), 105–121.

Ulrich, R. S., Simons, R. F., Losito, B. D., Fiorito, E., Miles, M. A., & Zelson, M. (1991). Stress recovery during exposure to natural and urban environments. *Journal of Environmental Psychology*, 11(3), 201–230. doi:https://doi.org/10.1016/S0272-4944(05)80184–80187.

Vega, K. A., & Küffer, C. (2021). Promoting wildflower biodiversity in dense and green cities: the important role of small vegetation patches. *Urban Forestry & Urban Greening*, 62, 127165. doi:https://doi.org/10.1016/j.ufug.2021.127165.

Vieira, J., Matos, P., Mexia, T., Silva, P., Lopes, N., Freitas, C., … Pinho, P. (2018). Green spaces are not all the same for the provision of air purification and climate regulation services: the case of urban parks. *Environmental Research*, 160, 306–313. doi:10.1016/j.envres.2017.10.006.

WHO. (2014). *Quantitative Risk Assessment of the Effects of Climate Change on Selected Causes of Death, 2030s and 2050s*. Geneva: World Health Organization.

Wilson, E. O. (1984). *Biophilia*. Cambridge, MA: Harvard University Press.

Wolch, J. R., Byrne, J., & Newell, J. P. (2014). Urban green space, public health, and environmental justice: the challenge of making cities 'just green enough'. *Landscape and Urban Planning*, 125, 234–244.

6
CO-PRODUCTION

Embracing Gaia

A former warehouse in Kolkata, which operated during the colonial era, lies invaded by stranglers. The roofing of the colonial structure is missing; leaving behind a skeleton of roof-trusses and shattered windows, as though the destroyed remains of a bombed town; and yet the red brick masonry stood covered with roots. The building still stands here, not as a monument of war but serving as a reminder of the colonial era. Others might associate it with a "failed future" of the British rule or a "destroyed place" through abandonment leading to what Edward Relph discusses as "placelessness" (Relph, 1976). However, Christopher Woodward in his book *In Ruins* published in 2001, point to how abandoned and derelict spaces can become places that are aesthetic, uplifting and comfortable (Woodward, 2001) – suggesting the subjective reality of such spaces as observers are forced to "supply the missing pieces from their own imagination" (Woodward, 2012, p. 17). The structure has a collapsible gate that is locked, but unlike other locked and open-air structures, there are no signs of wild growth inside; just fallen leaves shed by trees. It would not be surprising to find out that someone cleans the enclosed space from time to time since access is rather easy, and the building is a part of a campus that is still in use. Yet, the assemblage of abandoned warehouse stands as an island, with no purpose, amid an otherwise thriving life of the city.

DOI: 10.4324/9781003374602-6

The roof of the building seems to be missing – perhaps asbestos sheets which once covered it were dismantled and reused elsewhere. Despite light and rain to its inner courtyard trees have found it convenient to grow only on the walls. As if nature was determined to show its truest disregard for human control, and do what is forbidden. Such examples of nature regaining control of urban spaces due to human neglect perhaps represents the microcosms of James Lovelock's hypothesis of Gaia (Lovelock, 2000). James Lovelock's book *Gaia: A New Look at Life on Earth* has been reprinted several times since 1979, and in the 2000 edition the author improvises on previous flaws in his imaginary of our planet Earth as "a kind of living organism"(Lovelock, 2000, p. xv). As it stands, Gaia in the broad sense, is an understanding of Earth's restorative force that responds to the imbalances caused in the ecosystem. We notice the forces of Gaia at work, as nature appropriates human creations to co-produce an assemblage which is rare to see, and stimulates the mind to think about the future of urban areas. There has been at least some curiosity among scientists, as to what ecological succession of "urban ecosystem" would look like.[1] The natural model can be imposed, but not without the existing relics of human existence – the buildings, automobiles, roads, paved paths, streetlights, electric poles, and all other familiar structures which fuse with natural elements to create strange settings of built and nature.

However, a complete return to the natural should not be romanticised.[2] Because as mentioned earlier in Chapter 3 on Existence, in urban areas, when nature regains control, it doesn't return the system back to the original state; instead forms a novel landscape that integrates human-made elements within itself. While novel settings may emerge from human negligence and abandonment, Collier and Scott (2008) points to how humans do not always abandon spaces completely, and sometimes continue to establish new links with these sites.[3] This process is spontaneous and typically are outside the societal norm. Not to mention, they may have both positive and negative effects on society: for instance, abandoned structures and neglected parks that are out of sight of public may be spaces for promiscuous activities and gambling; on the other hand, people may also take advantage of the authority's oversight to experiment with graffiti as well as engaging with more playful and exploratory activities that encourage a stronger human bond with nature. The latter can be as simple as children balancing over fallen tree branches or being fascinated by colourful insects and plants that inhabit these spaces. Indeed, a wide range of contrasting outcomes are possible associated with wild spaces in cities. Notably, what the above also highlights, is a cyclic relationship between humans and nature – that, unlike in many natural ecosystems, ecological succession in cities is not linear in its transformation. Human and natural forces are going to interrupt each other's dominating streaks – transforming and re-transforming urban landscapes.

Having said that, eventual trajectories of wild urban spaces are to end up being recycled for human uses. One common approach is rebuilding the sites entirely. Since my visit to Kolkata, several colonial buildings with heavy tree invasion, which I had found intriguing have already been demolished, to make way for newer buildings with updated architecture. Particularly, the formerly industrial sites have piqued interest of modern cities that are now aspiring to convert these into commercial and cultural spaces to meet the requirements of economic change. Another trajectory of these sites is reappropriating or renovating them to adapt to current demands of aesthetics and use of space. This trajectory involves some amount of preservation. Think of decommissioned tramways, elevated light rail, or obsolete factories being converted to vibrant parks; or industrial land being recycled for constructing office buildings and housing. Memories held through relics of industrial legacies in these sites also make them ideal for being reused as museums, art galleries, and exhibition centres. The spaces are almost always reconfigured for new uses, often designed from scratch, wiping out any established links between human-made elements and wild growth in the sites. Yet, inevitably, this sets the clock ticking for the next cycle of failure of urban spaces that might lead to loss of human control over long periods of time, that will allow nature to grow again.

If plants can reclaim land, simply by adding to human-altered landscapes, why cannot humans mirror this aspect of nature and rethink the purposes of reclaimed landscape without removing or replacing the aesthetic compositions created by humans and nature? The process whereby the two parties – humans and plants – create a landscape together is what I refer to as *co-production*. In other words, for a landscape to be co-produced humans must make use of sites that nature has reclaimed, without destroying it – through the process of adaptation response and not a reset. The idea of co-production (often also termed co-creation) among social scientists and town planners, however, has been more to explain the various human stakeholders that come together to create solutions that "work". This human-centric usage of the term co-production, in urban studies and planning, usually flows from addressing a complex political ecology; or as Bruno Latour suggests "bring[ing] the sciences into democracy" (Latour, 2004, p. 221). However, in my case, such understanding is not enough to explain the temporal relationship between human and non-human agency. Thus, in borrowing the term "co-production", I digress from, and perhaps expand its usual meaning for, explaining the ages and stages of how nature and humans interact to accommodate each other's creations.

While so far, the theories about co-production have almost always used a medium or as Latour calls "intermediary" that is finally a human actor, taking the lens of Gaia, helps us imagine how in the case of landscape fractions, nature does not need an intermediary, but rather the lack of it. Co-production here is a response, that of course needs a voice that roots in favour of retaining the spontaneous botanical growth on buildings.

FIGURE 6.1 Premature stage of co-production: giant stranglers growing on top of an abandoned warehouse in Kolkata. March 31, 2018.

The reclamation of urban surfaces by plants can be thought of as only a premature form of co-production (Figure 6.1). At this stage, while the plants have adapted to the human-built surfaces, humans have not managed to adjust to the newly formed landscapes. The premature stage of co-production is an incidental occurrence; as would have been elaborated upon in Chapter 3 on Existence. Then it could have neither intended to destroy the landscapes, nor even beautify them, but simply must have turned out to be. A more mature form of co-production should find some sort of response from the human agency as well, that allow integration of reclaimed landscapes into the designing of (urban) spaces. Take, for example, simple acts such as the pruning of wild plants to control their growth or adding a pathway to create access through an otherwise wild grass-bed. Essentially, these acts allow humans to adapt and assimilate landscape fractions more purposefully in their lives.

Disappearance of landscapes

Spontaneous botanical growths have defied hostile environments in cities to survive where planned nature has disappeared – yet they are neglected without being integrated into urban spaces. Without human intervention to protect and enhance their purpose, these remaining assemblages of nature are at risk too. The risk of losing such landscape fractions is probably greater than losing nature in general, because as I have highlighted, such assemblages are often rare occurrences, which are formed by incidental interactions between human-made relics and spontaneous natural growth. Here, I emphasise the word – incidental. Under the circumstances, if the critical socio-political or physiological links to their existence are broken, these settings may not be recreated once lost. As a result, for some neighbourhoods it may even mean losing their only bit of nature.

Returning to the example of the park on the outskirts of Bengaluru (see Chapter 1 on Morphology); it is the only public park of the neighbourhood. The infrastructure is well equipped with non-living embellishments, but lacked efforts in maintenance of ornamental plants. Instead, one can find wild growth, not everywhere within the park, but in various strips and patches, taking over the human-made objects – climbers over the fencing mesh and entwined with the pole of the streetlight; mosses underneath the concrete seatings; flowering herbs that mostly aligned themselves behind the kerbs but at places spilled over to the walking track; a sea of shrubs and tall grasses around the metal water standpipe, and so on. The few mature trees existing on this site are very old, and in the process of having their life being sucked out by strangler plants growing on them like parasites. Some trees which were felled due to risks to human life, were never finally replaced. Wild nature in the park blends in quite well with the landscape, although is removed from time to time by public authorities to manage overgrowth. Each time the wild growth is trimmed or removed, there is an anticipation among the users that new designs are going to be implemented. But so far, efforts for maintaining designed landscaping have not been entirely successful in the long-term. The only future of the park would then seem to be the patches of spontaneous growth that have consistently put up a fierce battle against human control.

The temporal complexities of such assemblages should not be discounted. Interestingly, over the many years I observed the park, each time after wild growth was removed, nature grew back with a different character – sometimes dominated by wildflowers that drew butterflies, other times by tall grasses that provided habitat for fireflies. It is perhaps worth mentioning here, that the environmental conditions were also never constant. Sometimes, the wild growth was nurtured by the rains, other times left to mercy of a hot and dry summer. Either way, the neglect allowed spontaneous nature to become a mode of greening in the park (Figure 6.2). With its full bloom, the uncared or wild had its own charm. The strip of wild growth alongside the jogging tracks offered a burst of green for morning walkers, as well as, a bed of colourful flowers for those who choose to sit down and immerse themselves in nature. Partial management of the landscape is perhaps necessary to deal with uncontrolled growth that beyond a point hinders socially productive human activities in the area. But complete removal of this nature is quite palpably a loss of aesthetics and habitat – no more butterflies, and no more greens.

Small-scale growth of wild nature is seldom noticed as objects to preserve. On one hand there may be deliberate attempts to remove the wild growth of shrubs from the roadsides to make the neighbourhoods look cleaner. On the other, fleeting landscapes that had come into existence due to temporary stagnation of water disappear on their own when the sunny days return. Not to mention, some plants, despite finally surviving the harsh conditions, are ignored as expendable nature. Perhaps this problematic can also be viewed as an extension of Gaia: because if the

FIGURE 6.2 Gradients of managing spontaneous landscapes in a public park of Bengaluru: wild growth in its full bloom offers a unique aesthetic for visitors.

FIGURE 6.3 Removing ground cover: the verges of a street in Bengaluru which earlier served as grazing grounds and sponge spaces are now fully paved. The streetscape looks "clean" and "new", and saves on maintenance trading off one stratum of biodiversity. c. 2021–22.

restorative forces of the earth help restore equilibrium at some point, then why should it matter if the nature is lost? It will grow back again. The everyday landscape as well as abandoned spaces in which wild growth is so free to come to life and die, is living proof of hope that nature does not need human care to grow, and revegetation can happen in due course of time. However, it misses the main link, i.e., what links do these wild spaces have with human life in cities? It is helpful to remember that neglected spaces are not just abandoned industrial yards or locked dilapidating properties which are seldom used by people. They are also street-side verges and parks which are part of our everyday nature and is well accessible to people.

The disconnect between the top-down envisioning of spaces and the everyday reality is well visible in the following case of white topping implementation in Bengaluru. White topping is a city-wide decision by the public authorities, and refers to concreting of pavements or road surfaces. However, it is hard to understand what was imagined, and it is not known, if there has been any kind of discussion on developing street designs that accommodate for more nature in cities. It is also partly how designs are implemented that is problematic. For instance, sometimes, no formwork is used to protect spaces around trees or verges. The slurry is simply dumped on the right-of-way, to fill up the road space. Such wide-scale urban planning decisions have serious implications on people's lifestyles, as the street verges would no longer support plant growth to serve as a grazing network for cattle (Figure 6.3). In cities the pressure from human design can change landscapes so rapidly, and with such long-lasting effects, that despite the restorative property of nature, wilderness in most cases does not have a fighting chance. These linkages are important to bring to the consciousness of people, as citizens may not realise the value of wild spaces even after they are lost – the renovation of road that led to loss of habitat and grazing grounds for livestock goes remiss, as neighbourhood dwellers are swayed by the new aesthetic.

In other cases, as I have explained in Chapter 4 on Aesthetics, the assemblages draw their meanings from the historical background of the involved elements; which embodies their purpose in some time-period or context. It is not uncommon to find sudden sprouting of green growth on buildings (and other corners of the city) only in certain seasons of the year, and disappear after their seasonal lifecycle is completed. These landscapes, although resilient in some sense, can also be sensitive to certain changes in their environment. Thus, it serves good purpose to discuss a few examples to explore how landscape fractions have sometimes survived epochs of interruptions in their trajectories by human and natural agencies. This is not just to explain the forces through which assemblages happen to exist (as done in Chapter 3 on Existence), but instead to understand in particular, the processes in which human agencies re-engage with spontaneous nature in cities to create harmony between different elements to move beyond mere existence and co-produce an aesthetic.

Harmony – an idea and its processes

In cities, non-intrusion by humans has allowed nature to thrive, retaining its own character – thereby leading to the creation of novel assemblages. However, the lack of initiative to intervene also prevents adding value to nature. To add value to nature, in a manner that synchronises the need of humans with the formations of reclaimed landscapes – would be co-production. Such synchronisation typically happens in three waves of production: 1) natural landscapes are transformed by humans; 2) nature reclaims human-made surfaces; 3) reclaimed spaces of nature are assimilated as part of urban spaces by humans through designs that respond to these reclaimed precincts. To illustrate this process one can, picture epochs of urban fabric transformation as follows: first, the cities have been created by clearing natural landscapes; second, spontaneous growth has appeared over this human-built urban layer; and third, humans have the choice to now remove these configurations of nature and rebuild, or they can design environments that respect these human-nature assemblages which have been created.

Landscape transformations can be either rapid or can happen slowly over time. Take, for instance, a slow transformation of the landscape caused by the night-time parking of bulldozers and trucks in the playground. Their heavy weight and giant wheels create depressions, shaping the terrain for retaining water and creating shallow pools and islands – eventually turning the ecology of perfectly dry land into that of a wetland. Examples of more rapid changes can be traces of human activities within a park that has wild growth. A pack of cards from illegal gambling events, crushed plastic water bottles from thirsty visitors, and paper cups that were bought from vendors selling tea are relics that can be commonly found. Less common are automobiles and large technological pieces of equipment. Sometimes such changes create interesting assemblages by combining with human-made relics and other times destroy aesthetics of the spontaneous landscapes through littering. More on how such notions get constructed is explored in the next chapter on Frames.

Noting how urban landscapes are similar yet not same as the rural landscapes, Finnish botanist, Jari Nimelä lays out three broad steps for integration (Niemelä, 1999):

- Step 1 – to find different types of nature in cities
- Step 2 – to acquire knowledge about urban nature
- Step 3 – to design urban nature based on acquired knowledge

These steps also apply to small-scale spontaneous nature in cities. Step 1 is about identifying, discovering, and documenting assemblages which are discussed more in Chapter 1 on Morphology and Chapter 7 on Frames. Step 2 encompasses both the scientific and socially constructed meanings of landscape fractions that are discussed in Chapter 2 on Meanings and Chapter 5 on Collectives. This chapter partly addresses some of processes by acknowledging the oscillations of power between natural and human agencies in shaping these landscapes as well as to suggest somewhat of a way forward. It will also inform step 3, by highlighting a few potential ways in which particularly small-scale assemblages of spontaneous nature can be integrated into the socio-material urban fabric by allowing a stronger relationship between people and nature.

Past research on larger patches of spontaneous nature has already indicated that it is possible to integrate wild landscapes in urban areas for people's use through design interventions. Del Tredici (2010) summarises the contributions of various authors in Kowarik and Körner's edited book[4] (published in 2005) into three main areas:

- People's preferences matter in managing such spaces
- Manipulation of landscapes can enhance aesthetic qualities
- Retrofitting of human-built landscape features—such as pathways and art installations can encourage more use

Particularly, research has found that small interventions into spaces of spontaneous nature can encourage marginalised groups. Studying a harbour area of Tallinn in Estonia, Unt and Bell (2014) documented how small improvements in an otherwise derelict urban site increased visits of women and older people by five times.

I briefly draw attention to how cities have adopted an "ecological mode of management" (Ernwein, 2020, p. 10), creating interesting more-than-human assemblages, where wild growth is managed through a less intensive approach – often meaning a return to traditional techniques more suited for agriculture rather than the urban spaces – that avoids over-management of urban green spaces. For instance, use of scythe instead of lawnmowers in Geneva (Ernwein, 2020). Or for instance, how, in the Netherlands, there is paid grazing where sheep are used as lawn mowers in city parks and other green spaces in the public realm that are growing completely wild.[5] Notably, both approaches offer organised sources of livelihood in the formal sector. One Dutch citizen explained "shepherds make a living on that – they have vans, they have their dogs, electrified plastic fences, and sheep. When they get an assignment, they set up their fences in the space, and let their sheep graze".[6]

The above insights are drawn for relatively large land parcels within cities. From here on in this chapter I will demonstrate, with a few examples, how the aforementioned findings are applicable for small-scale nature.

Pruning or trimming wild landscapes has been one way to reduce risks to structures and manage overgrown nature. Plant growth on the boundary wall in Hrishikesh Laha's compound on Amherst Street in Kolkata, is pruned periodically by the decision of landowner, but the roots remain for the plants to grow back. Removing the roots would expose the masonry and thereby damaging the structure even further. This notwithstanding the overall safety of the structure should be monitored as a public road abuts it. As another example, I return to my site – the suburban neighbourhood of Bengaluru, where often, wild shrubs on the street-side verges are found to overgrow, and block entries to the houses in front (Figure 6.4). When this happens, homeowners hire gardeners to trim the overgrowth – just enough to clear the entries.

Programming compatible activities around the spaces unlock the potential of novel assemblages. While plant growth on buildings is largely seen as a threat to structure and sign of urban decay, these settings are comparable to what Jorgensen and Tylecote (2007) describes as ambivalent landscapes, i.e., they simultaneously hold positive and negative meanings. Biologists point out how the eventualities related to

FIGURE 6.4 Community management of wild spaces: a privately hired gardener trims an overgrown patch of wild plants that had blocked the entry of a residence in Bengaluru.

building decay, especially related to adverse effects on human health stops one from arguing habitat benefits of such sites. At the same time, dilapidating buildings have drawn attention from film makers who have been able to use shoot horror films in such settings. Another example is how the soap factory in Bengaluru now offers occasional guided tours to the public inside the old factory; giving access to a variety of new and interesting assemblages that can rarely be found elsewhere within the city. For instance, the effluent treatment tanks inside the factory have created microhabitats for fish, tadpoles, water striders, dragonflies, and algae (Figure 6.5). These settings can be useful for educational excursions as opposed to being locked away from the public entirely. They offer an interaction with nature, of course with the necessary safety measures in place to do so.

In earlier chapters it has been highlighted, through different examples, how ecological habitats are often created in places which are locked away from, or are restricted to, the public. For instance, the effluent

FIGURE 6.5 Biofilm in an effluent treatment tank within the soap factory of Bengaluru: this setting has offered ecologically stable conditions for growth of algae and become habitat for fish, and insects. January 14, 2020.

treatment plant which is used to treat water discharged from the soap factory in Bengaluru – has the right chemical properties to give rise to, and support, a host of living organisms. The sedimentation tanks of the treatment plant now serve as pools, but unlike the waterlogged playground, has been able to offer more stable conditions for biological productivity. The tank becomes an interesting site with live organisms creating ripples in the water or hovering over it – both of which can be appreciated by the naked eye. They encourage the observer to come a little closer and observe the actions of fish and tadpoles that swim in and out from under the varied shapes of floating algae – thus making the tank a site of ecological interest.

Resolving contestation for land

On a bright sunny day, I walked through the outer ward of the seventh-century heritage site of Whitby Abbey. The remains of the medieval site maintained by English Heritage consists of the stone buildings of the castle (also known as Dracula's Castle), the abbey, and the walls that enclose them. Within the walls is a small cemetery, a pond, and vast expanse of grasslands. Walking for long enough, it would take one out of the castle's perimeter, down from the hill, and connect with roads that finally lead towards the city centre along the River Esk. The path was laden with flowers growing on either side amid the tall grasses (Figure 6.6). The tall grasses are representative of what would have originally been farmland. Indeed, agriculture and animal husbandry were a typical element associated with Abbeys to ensure self-sufficiency of the monastic communities. Incidentally, the grasses which I now presume were maintained by the English Heritage, have become home to birds, and what a delight it was to find flightpaths of birds criss-crossing, over and under, the grasses of this landscape. The metaled road is invaded with ground cover. The built tarmac road can become a metaphor for human efforts that aimed to keep nature away from human footsteps. However, with time, patches of wild plants have grown over the road and nature has clearly begun reclaiming its land. Yet, this invasion is harmless to the road's actual utility of offering a comfortable and designated walkway for the pedestrians. Thus, metaphorically speaking again, nature has been sensitive to human-use of the site. Rather interestingly, these patches of microsites have become islands of biological diversity. As I followed this road, I found patches of ground cover growing denser. A mosaic made of the patterns of wild growth (Figure 6.7), should show how a diverse range of botanical species have managed to conquer the road space: with a variety of tiny leaf structures and tiny flowers, as if walking on stars beneath my feet. Towards the end of the path that led to the outside of the the walled area, grasses had completely taken over the road, and signs of trimming hinted to the efforts of landscape managers having mowed the grass to designate the walkway. I noticed how the plants were not uprooted, nor the road surface had been artificially sealed to stop this botanical invasion. The grasses are so finely ingrained in the road, that any means adopted to uproot them completely will have to cause damage to the road itself. Rebuilding or repairing the road is of little extra use, and keeping the nature intact is of no harm to anyone when the path can still be used while having a green-aesthetic touch to it. This may not be a sign of neglect but a choice to preserve and enhance the heritage value of the medieval site.

The benefit of small-scale nature, as from this example, is their ability to invade human-made surfaces without hampering the utility of the

FIGURE 6.6 Nature integrating with a road within Whitby Abbey grounds. June 3, 2017.

structures to which these surfaces belong. These landscapes tell a story in sequence. A story of how nature reclaims land by invading human-made structures and furthermore – how humans respond in return. In this case, the human response has had to preserve nature's reclamation creating a unique setting. The landscape as part of Dracula's castle in Whitby is perhaps a one-stop exhibit to demonstrate the different stages of how humans can assimilate nature reclaimed construction. Fractions within the site serves as a miniature setting to demonstrate urban processes and Gaia. The site is an embodiment of how people make a conscious choice to retain nature to enhance heritage of urban spaces, and how not all nature over built surfaces is harmful.

Humans need space to build, and nature needs space to grow. This is one of the fundamental reasons behind our conflict with wild nature. As per traditional trajectory, for cities to support their growing populations, more nature is consumed – and transformed either into built environment or agricultural land. The scale of our global operation to replace nature is so huge that it might be hard for one to picture the change. If each year, global forest loss is the size of the UK (Harvey, 2019), then

FIGURE 6.7 Competition for space: mosaics formed as a variety of herbaceous plants (or ground cover) are gradually taking over a road in Whitby Abbey. June 3, 2017.

that scale of change is not easily perceived with our eyes and are, instead, a combination of snapshot visuals and satellite imageries that can represent this change. Moreover, environmental changes across the planet have become so intense that they are often perceived to have reached a point of no return. That is why ecological succession in cities is so interesting. They may hold abstract models and more perceivable, human scale, everyday experiences of what it would be like for nature to take back control of Earth's surfaces. I find wild spaces within cities as a microcosm of what the Anthropocene is doing to nature and nature's response to reconquer its lost territories. So far, I have discussed how people manage landscapes through programming new activities and through managing spontaneous botanical growth as a mode of greening. However, to visualise the stages of development, it helps if one witnesses the transformation of the sites over time. Perhaps there was no better an example which I found during my years of fieldwork that highlighted a gradient of human-nature relationships that exhibited the premature and mature stages of co-production side by side, as much as Whitby Abbey.

Creating small urban places

Small precincts of nature can create wonderful places. They might well be afterthoughts. To support this argument, I do not draw from planned pocket parks of a modern part of the town but find a precedent in the porches of colonial-age buildings. Comprised of narrow streets in the traditional core of Kolkata, these neighbourhoods are typically devoid of any nature. But one street somehow has it very different. House owners have built small niches and planters which allows not just shrubs but also trees to grow (Figure 6.8). These plants are not spontaneous nature – because they are planned – however, it is not a complete product of premeditated design either – as retrofits to the buildings to attain this form itself is a spontaneous process. The spontaneity lies in responsive design to a desire for reuniting with plants against the general trend of living in non-green neighbourhoods. The minimality has rendered such a street unforgettable to the passer-by who would benefit from the familiarity needed to navigate the intricate network of streets to carry out the routine chores of everyday life. To a visitor, however, this is no less than a secret place awaiting discovery. Then it is certainly intertwined with emotions of surprise and relief to find even small-scale nature in otherwise completely grey neighbourhoods.

FIGURE 6.8 Exemplar precedent of small-urban places: even if lacking space, residents have grown plants inside planter boxes in Kolkata. July 15, 2018.

These arrangements reveal that plants can be accommodated in the extreme scarcity of space in built environments. More importantly, the residents have chosen not to maintain manicured and rigid forms of landscapes like topiary and mowed grass strips – allowing the plants to grow naturally. Here stimulating shapes are not created through prefabricated ideas but organically by humans and nature. But there are elements that are missing from this being a residential neighbourhood; making it a quaint environment for private purposes but limiting the potentials of public placemaking. What is missing? This I analyse through components for placemaking highlighted by Sherman (1988) adapted in Montgomery (1998) which I chart in Table 1. And it is pleasurable to imagine the same street with commercial fronts which offer food and drink, comfortable places to sit and wait, the opportunity for idle interaction between people that heighten their emotions and add a sense of belonging; serving the potential of the place to be far more experimental.

Small urban places need not only be in the public realm and can also be created in private properties in response to spontaneous nature. For

TABLE 6.1 Evaluating the exemplar precedent as an urban place

Success factors of urban places	*Evaluation*
Look natural	✓
Stimulating shapes	✓
Familiarity with streets	✓
Secret places for exploration	✓
Surprises for conversations	✓
Experiments and excitement	✓
Opportunity for informal casual meetings	x
Food and drink	x
Comfortable places to sit and wait	x
Balance between loneliness and privacy	x
Celebrate changing seasons	✓
Heightening senses (though verbal and visual stimulation)	✓

Based on Sherman (1988) adapted in Montgomery (1998)

instance, consider Figure 6.9 which is a front yard of a residence, whose residents have not opted for plastering or painting of walls or even planting and maintaining horticultural plants – keeping expenditures low. The narrow path, however, created by laying bricks, not only offers continuity from the exposed masonry of the walls but also adds definition to the spontaneous growth around it – creating a harmonious landscape composition. The central idea remains co-production or retrofitting without a complete redesign (e.g., paving which saves on costs of landscaping). The vistas, however, allow the aesthetics to be enjoyed by the public when passing by the compound gateway. When these vistas form part of neighbourhood fabrics with narrow streets and very little public realm, they offer bursts of green visuals as one travels around an otherwise dense and built-up locale.

To an extent, examples in this chapter, of managing spontaneous landscapes, may be considered as "hybrid landscapes",[7] which speaks to both top-down allocation production of space, and community-led small-scale interventions (Quayle & Lieck, 1997). Indeed, wild plants may emerge on formal spaces (such as roadside verges, front yards, and courtyards) otherwise allocated for designed nature; and thereafter managed by residents themselves. Karvonen and Yocom (2011) argue

FIGURE 6.9 Bricks laid to form a path over wild vegetation in a front yard of a residence in Kolkata. April 14, 2018.

that the prevalent "top-down" and "expert-driven" approaches to managing nature simplify the human-nature relationships. Hybrid landscapes allow more interaction with nature which is argued to elevate people's connectedness to nature, which means people will also be more likely to protect plants when at risk of being lost by redevelopment activities (Quayle & Lieck, 1997). Having said that, in Bengaluru, road constructions wiped out roadside spontaneous nature, managed by communities – without facing any protest. Perhaps because spontaneous nature is not managed with the intention of retaining them but it just happens to be so. Thus, to expect a civic movement in favour of such nature, a certain amount of awareness is still needed to help people realise the value of what they are losing.

A little maintenance goes a long way. When sites are vandalised or covered in litter, even the most interesting sculptures of nature and humans can look bad. In Figure 6.10, I exhibit two sites: site A has unmanaged growth, a broken control box with worn-out paint, and is surrounded by leaf and plastic litter. The control box in site B on the

FIGURE 6.10 Comparing maintenance levels of two sites in Bengaluru: site A has a broken control box with litter around it; site B has a control box that is intact with wild flowers growing around it. c. May–June 2020.

other hand is in good condition; and colour combination (cool shades of green and red) of the paint creates balance. In addition, there is the right amount of plant growth controlled by built surfaces around the control box, and the tiny yellow flowers of the straggler daisy (*Calyptocarpus vialis*), the otherwise rampant invasive species in the city, here has become desirable for creating harmony. Site B exhibits the right ingredients for managing small-scale wild spaces as landscape fractions.

Notes

1 Zipperer (2010) reviews the literature to discuss three key aspects of urban ecological succession: availability of land, availability of plant species, and performance of plant species.
2 Drawing from literature review, Perring, et al. (2011) notes that natural and social barriers tend to prevent novel ecosystems from returning to their historical states.
3 See also Collier (2015).
4 Cf. Kowarik, Ingo, and Stefan Körner, eds. 2005. *Wild Urban Woodlands*. Berlin: Springer.
5 Informal discussion with a Dutch citizen on urban grazing in the Netherlands, 2022.
6 Informal discussion with a Dutch citizen on urban grazing in the Netherlands, 2022.
7 Hybrid landscapes are co-produced together with advisors (e.g., street engineers, park managers) and the residents, to both of whom municipal powers need to be transferred (Quayle & Lieck, 1997).

References

Collier, M. (2015). Novel ecosystems and social-ecological resilience. *Landscape Ecology*, 30(8), 1363–1369. doi:10.1007/s10980-015-0243-z.

Collier, M. J., & Scott, M. J. (2008). Industrially harvested peatlands and after-use potential: understanding local Stakeholder narratives and landscape preferences. *Landscape Research*, 33(4), 439–460. doi:10.1080/01426390801948406.

Del Tredici, P. (2010). Spontaneous urban vegetation: reflections of change in a globalized world. *Nature and Culture*, 5(3), 299–315.

Ernwein, M. (2020). Bringing urban parks to life: the more-than-human politics of urban ecological work. *Annals of the American Association of Geographers*. doi:10.1080/24694452.2020.1773230.

Harvey, F. (2019, September 12). *World losing area of forest the size of the UK each year, report finds*. Retrieved January 27, 2023, from *The Guardian*: https://www.theguardian.com/environment/2019/sep/12/deforestation-world-losing-area-forest-size-of-uk-each-year-report-finds.

Jorgensen, A., & Tylecote, M. (2007, August). Ambivalent landscapes – wilderness in the urban interstices. *Landscape Research*, 32(4), 443–462.

Karvonen, A., & Yocom, K. (2011). The civics of urban nature: enacting hybrid landscapes. *Environment and Planning A*, 43(6), 1305–1322.

Latour, B. (2004). *Politics of Nature: How to Bring the Sciences into Democracy*. London: Harvard University Press.

Lovelock, J. (2000). *Gaia: A New Look at Life on Earth*. New York: Oxford University Press.

Montgomery, J. (1998). Making a city: urbanity, vitality and urban design. *Journal of Urban Design*, 3(1), 93–116.

Niemelä, J. (1999). Ecology and urban planning. *Biodiversity and Conservation*, 119–131.

Perring, M. P., Manning, P., Hobbs, R. J., Lugo, A. E., Ramalho, C. E., & Standish, R. J. (2011). Novel urban ecosystems and ecosystem services. In R. J. Hobbs, E. S. Higgs, & C. M. Hall (Eds.), *Novel Ecosystems: Intervening in the New Ecological World Order* (pp. 310–325). West Sussex: Wiley-Blackwell.

Quayle, M., & Lieck, T. C. (1997). Growing community: a case for hybrid landscapes. *Landscape and Urban Planning*, 39(2–3), 99–107. doi:https://doi.org/10.1016/S0169-2046(97)00048–0

Relph, E. C. (1976). *Place and Placelessness*. London: Pion Limited.

Sherman, B. (1988). *Cities Fit to Live In*. London: Channel 4 Books.

Unt, A.-L., & Bell, S. (2014). The impact of small-scale design interventions on the behaviour patterns of the users of an urban wasteland. *Urban Forestry & Urban Greening*, 13(1), 121–135. doi:10.1016/j.ufug.2013.10.008.

Woodward, C. (2001). *In Ruins* (1st American ed.). New York: Pantheon Books.

Woodward, C. (2012). Learning from Detroit or 'the wrong kind of ruins'. In A. Jorgensen, & R. Keenan (Eds.), *Urban Wildscapes* (pp. 17–32). New York: Routledge.

Zipperer, W. C. (2010). The process of natural succession in urban areas. In I. Douglas, D. Goode, M. Houck, & R. Wang (Eds.), *The Routledge Handbook of Urban Ecology* (pp. 187–197). London: Routledge.

7

FRAMES

Documenting assemblages

By now, the book must have established that small-scale assemblages of nature sometimes have values which remain unrecognised. This calls for a process of evaluation, where documentation is needed for assessing benefits of small-scale landscape assemblages. Of course, the reasons as to why different people would want to document plants would generally vary. As British photographer Heather Angel puts it in case of photographing trees:

> *The motivation for tree photography will often develop from a close working association with trees, and, for this reason trees will mean quite different things to the arboriculturist, the forester, the botanist, the landscape architect, the ecologist, the conservationist and the naturalist.*
> *(Angel, 1980, p. 105)*

Therefore, the values of assemblages are likely to emerge from differences in motivations of those that survey them. Tensions are inevitable: spontaneous growth which may become sites of biodiversity may also be the cause structural damage to buildings or clogging water flow in a storm water drain; aesthetic assemblages to pedestrians may be a nuisance to the plot owner; or even a health hazard from the point of view of city forestry and public health departments.

DOI: 10.4324/9781003374602-7

Visual and technical documentation of spontaneous nature assemblages serve a few important purposes towards determining their values. First, photography helps capture aesthetics of counter-utilitarian growth before they can be removed – thus in some sense, becoming a useful tool for conserving memory values. Second, ensembles of nature studied in this book are also incidental, and therefore, are not always permanent. Snodgrass and McIntyre (2011) write how natural landscapes can be subject to long spells of unattractive state.[1] So, documentation also helps capture their brief aesthetic existences before they disappear after completing their natural life-cycles. Third, recording detailed descriptions of their compositions through more technical devices (such as structural cross-sections and species documentation) can prepare these assemblages for investigation by city designers, scientists, architects, and engineers who can then decide if retaining these assemblages are worthwhile and feasible. One may also consider imitating benefits of such compositions, while eliminating, by design, threats to human or structural safety.

Photographic documentation can help people to recognise landscape fractions that exist in the city, so they can be elevated to the value of artefacts. Thoughtful visual compositions can bring to the fore an otherwise subdued character of the fractions, which excite both the eye and the mind. The "static" nature of architecture and plants makes it easy for aspiring artists and photographers to select their subjects and experiment with lighting and composition.[2] However, in order to identify values, the researcher must also be aware of the extended discourses around the elements that make up the composition. If an assemblage is a composition of parts, then the same plant growing on two different surfaces in our case can create two different compositions, if they are associated with different site-identities.[3]

In Chapter 1, I explained that the definition of a landscape fraction is typically not the entire scenery visible to the eye, but a "landscape within a landscape". This means that it is only a portion of a terrain that has undergone what Gandy terms as "sensory enframement" (Gandy, 2016). These sensory enframements are a construct of what is in the visual frame of the human eye. Now, imagine the twisted trunk of the fig tree that rises up the pillar in Hrishikesh Laha's compound.[4] The eye naturally moves up the organic lines formed by the roots to halt at the crown of the tree, thus adding movement to an otherwise static landscape. To begin with, one can think of "framing" in a visual sense, i.e., as borders like that around a painting that sets the limits to what the eye sees, recognises, and appreciates. This concept of "frame" can be extended to a real-time arrangement of objects which draws the eye to focus on certain elements of the composition, while blurring the others. While the eye notices the landscape, the mind agrees to focus its gaze on a portion of that landscape as if it was the whole.

The analysis of a frame (or the process of framing) is partly situated in the visual composition of an assemblage and partly how the assemblage is situated in the wider context of space and discourses.[5] While the entire process is that of appreciation, visual appreciation is more to do with seeing and contextualisation is more to do with interpreting. Discourses emerge from different components in the scene which serve as visual cues. For instance, a painting of an assemblage may appear beautiful because of the artist's skill in isolating the subject from its less-appealing surroundings.[6] If, in reality, the same assemblage is found in a downtrodden area of the city, it is more likely to be associated with decay (see Chapter 2 on Meanings). Such contextualisation is as much visual as it is notional.[7] Generalisations persist in people's minds about cities to be exciting or boring; for neighbourhoods to be affluent or poor; or communities to be inviting or aggressive. These are wider contexts within which our assemblages of nature are situated. The context is certain to affect how people perceive the value of assemblages.

In this chapter, my intention is not to explore in depth the rules or mathematical principles such as *golden proportion* or *rule of the thirds* which lead to beautiful visuals – those are aesthetic properties. Nor is it to suggest an empirical base for assessing what can or cannot be considered a desirable frame – these would inevitably be cognitive to the onlookers of landscape fractions, and perhaps a matter of further, more experimental research. Instead, my effort is to simply highlight a few examples of how components and contexts in different scenes contribute to composing the visual and narrative frame within which assemblages of interest may be situated. The task, however, is not as simple as drawing a box around an ensemble that has botanical growth and architectural elements, but to recognise the principles which have guided different elements within a landscape to form a composition, that can be held to the status of a landscape fraction.

Foreground and background

Framing of nature in cities can happen through architectural elements (e.g., grills, walls, windows, pillars), and urban design elements (e.g., enclosures creating vantage point for vistas, paving, electric poles, narrow roads, or use of public space and character of buildings). For instance, in Figure 7.1 a ribbon of green growth has cropped up at the bottom of a boundary wall of a residence. The public standpipe with running water in this image is an important structural part of the assemblage. Despite a concrete plastering around the tap, seepage of water has occurred from giant cracks in its base – this is evident from the damp bricks on which the plant growth has happened. Not all surfaces are conducive for plant growth. In this case, the water is absorbed by the wall and finds its way into the atmosphere through the exposed masonry – creating a micro-habitat. Thus, sandwiched between the concrete and the plastered wall, the strip of exposed masonry has created conditions for plants, and is acting as a trellis that guides plant growth. In all of this, the public standpipe is not only the source of the water, but also serves as a strong visual focal point. But a faded maroon tint of the smoothened wall surface behind, creates contrast, yet soft enough to keep a harmonious emphasis on the mellow but jagged green strands of plants. While spontaneous nature makes for an organic pattern, the

FIGURE 7.1 Framed between textures and planes: vegetation growing near a public standpipe in a street corner of Kolkata. April 14, 2018.

rigidity of human-made objects offers a stable structure to the frame either by directing the growth of plants or by drawing focus to them.

Architecture and enclosures within urban environments offer interesting opportunities for framing nature. Returning to Ramdulal Sircar Street in old Kolkata – being well-connected to a thriving local market just minutes away from the site, it is a busy place. Upon it, lies our ghosted building locked and dilapidating, whose walls are stripped of plaster and exposing its classic red brick masonry – a character that is prominent of old buildings going through natural wear and tear in this part of town (see Chapter 2 on Meanings). Plant growth has emerged on the top floor, inside a room with a broken roof. The broken roof turns the room into a void, which has the walls and traces of what could be presumed to have been a tiled roof, creating the sense of enclosure within which the plant growth has occurred. In addition, the balcony also serves as a giant planter box which allows stranglers and ivy to grow. Architects and engineers would appreciate the wooden trusses and layers of construction mechanisms that were used traditionally to build grand structures of its time, albeit for conventional and everyday use. The many extrusions on the surface of the building façade generate depth in the visual.

Indeed, an interesting frame can elevate the value of a landscape. It is enjoyable to view a landscape through different shapes and sizes of visual enclosures – which may as well be completely unrelated to the purpose of landscapes. Often, they come in the form of windows, grills, a sequence of columns, alleys, and other settings which offer a *repoussoir*[8] for the observer to appreciate not the whole landscape but only a part of it.

In the other way around, plant growth can also act to facilitate *repoussoir* to buildings (Figure 7.2). In fact, landscape architects have long used plants to frame views or create "windows" in the foreground that generates a composition by "locking" interesting elements within. Catherine Dee writes:

> *Vegetation can be arranged to frame views of more distant landscapes or landscape components, particularly foci. Tree canopies or vegetation trained on pergolas form overhead or complete frames. The complex and soft texture of foliage and branches "locks" sky and distance into the immediate "space" of the viewer, thus bringing them closer, mediating and giving scale to distant forms.*
>
> *(Dee, 2001, p. 190)*

FIGURE 7.2 Plants acting as repoussoir: stranglers grow on either side of a building façade in Kolkata. July 9, 2018.

Take, for instance, the precinct of an old house in Kolkata where plants draw cynosure to the built elements by screening out other nearby buildings. By itself, the building would be boring – with dull worn-out paint of what would seem to have been a tint of yellow, now turned pale and muddy. The slowly warping façade and crumbling cornices are signs of dilapidation due to a harsh local climate. The thick white canvas cloths put up on the balcony to stop the rain from entering are also giving way. The rusted grills and the exposed woodworks on the railings and louvres are tell-tale signs of oversight in maintenance. Yet the building elements are in harmony, considering the various shades of brown. Though a building is old, it is only as interesting in its aesthetics as any other building from the colonial era in Kolkata. Then why did it catch my attention? I thought the figs – growing suspended on either side – gave the building a neatly bound frame, as though two bastions supported it, thereby giving this assemblage a sense of stability. The broad and rounded leaves of the fig are a different accent to the other-wise intricate façade that is filled with straight lines, but matches the dense oval ornamental works of the grill. The plants accentuate the façade by protruding out – visually pushing the building to the background and therefore recomposing the scene. Disconnected from the ground, and subject to il-logics, the botanical growths should look out of place, yet certain properties make them appear to be at rest. The logic of photo-sensitivity and dangling of roots below makes it rather simple to imagine them as rising from ground up. The vertical movement of these two plants puts a halt to the flow of horizontal lines running from side-to-side as cornices, beams, louvres, and railings. Like any spontaneous growths that happen in different points of time, the plant growths within this fraction could have been unequal, but incidentally, to a pedestrian, they happen to look very even. The well-balanced visual bastions control eye movement to offer a visual composure and define limits to the frame. As such, this is a quintessential example of how a plant growth can frame the built elements to form a composite landscape. Such plant growth can also occur over walls, fences, or other smaller architectural

FIGURE 7.3 Backdrop in contrast with landscape: a variety of plants grow in front of a maroon-painted wall in Kolkata. April 14, 2018.

embellishments to have the same effect. I will revisit related risks in later sections of this chapter.

The background can also become an essential component in framing a fraction. However, when focusing on small-scale assemblages, one can't always expect to find a sky or clouds[9] to serve as horizon(tal) line or backdrop. These assemblages define more of a sculpture than a typical scene. Sometimes the backdrop is the sky but other times it may be parts of machines, paving materials, façades of buildings, and so on to which the plants have fused visually to form the aesthetic assemblage. In Figure 7.3, the maroon wall offers a contrast to green growth yet matching somewhat to the greenish-purple–coloured leaves of other plants creating visual cues for the observer. The background brings out prominently textures of different plants. Consequently, the vitality of the fraction has united some elements in the composition of the overall streetscape while completely blurring out others – framing a landscape within a landscape.

The fraction is far from static. Even a gentle wind can cause the row of plants to sway and dance creating motion and drama. The grass sways in a different rhythm than the taller sturdier shrubs. In this fraction, the plants are not fused with the wall, but have grown out of the cracks at its plinth: between the kerb stones that are meant to separate the road from the boundary wall. Together the wall, the kerb stones, and the extremely narrow space in between where the plants have grown, becomes one joint ensemble that is worthy of calling a landscape fraction.

But character of the plants will change with time, and thus, temporality becomes important because: a) if the plants dry out or die, then the assemblages would cease to exist, only to expose the bare and decaying architectural elements; or b) if the plants grow too big, then it can pose a risk to structural safety of the façades. Even if overgrowth is non-threatening, it is unproductive because it would cover the entire assemblage by leaf growth that hides interesting architectural elements. To reiterate, the background and foreground are both essential elements of the frame, and both living and non-living components are subject to change with time. The statuses of the assemblages as landscape fractions lies somewhere within this time frame, and are not always perennial.

Wider context of beautiful assemblages

Plants growing on ornate walls are sculptures in cities, so are islands of wildflowers and grass growing in neglected spaces and residual spaces.

While our eyes particularly focus on assemblages of spontaneous nature, the surrounding context does not escape and instead creates matter for discourse. Discourses determine if these compositions belong to the heritage of the city or are more representative of its decay and disintegration of the social fabric. The wider context or scene becomes important as it offers elements that construct discourses through those who visit our sites of landscape fractions. Before proceeding further, I draw upon a translated excerpt from sixteenth-century agronomist Davanzatti, which reflects how boundaries of knowledge guide our value judgements:

> *From here below, we have difficulty in perceiving the few things that surround us, and we give a price to them according to whether we perceive them to be more or less in demand in each place and at each time. The merchants are promptly and very well advised of these things, and that is why they have an admirable knowledge of the price of things.*[10]

Of course, I might have served these lines out of their context. But to think, it is indeed part of reality, that some people would be able to appreciate beauty more instantly than others. Then "merchants" who are well aware of values, in our case might be architects and designers who, through their training and experience in beautification and placemaking, must know the worth of these assemblages. Landscape fractions are a useful tool for city design as the value of an absolutely stunning assemblage cannot be dismissed.

Today, it is relatively well regarded than before that growing cities need to accommodate for more nature. Health benefits associated with greening make urban centres liveable (Davern, et al., 2017). Yet per capita green space in cities is low, particularly in low-income neighbourhoods. Given this context, people can perhaps be expected to grow more tolerant of wild growth, even if not turn entirely in favour of them. However, negative influences from the wider landscape are not uncommon and prevent landscape fractions from attaining their full value in performing roles such as landmarks and identities of public places.

Risk to life

Depending on their structural context, assemblages of spontaneous nature, by their virtue of unmanaged character can pose risks to life as well as human health and wellbeing. To begin with, the most serious risks come from plants growing on old buildings that might lead to structural damage by dislocating masonry (Figure 7.4). Practitioners and scientists hold that the additional load and moisture retention by plants on buildings can lead to the structure's collapse, and that such risks of injury to the occupants prevent them from looking at these sites in the light of their other advantages such as aesthetics or ecological habitats.

Certainly, the risk to people's lives is non-negotiable. This strong disincentive also stops biologists and others who are interested in plant growth on buildings from expressing their enthusiasm of plant growth, even in relatively less risky conditions: say on abandoned buildings and non-structural walls. This notwithstanding, abandoned buildings have had the problem of rainwater stagnation – sometimes aided by plants, which pose a risk of waterborne diseases such as malaria and chickunghunia.

FIGURE 7.4 Context as risk: trees growing on the roof of buildings within a residential neighbourhood in Kolkata. Some buildings here are inhabited, while others are vacant. July 9, 2018.

Yet sometimes, to safeguard historic buildings, plant growth has to be retained. Citing the example of Ta Prohm, a twelfth-century temple in Cambodia, one conservation architect highlighted how "the plants have now become an integral part of the building, and any attempt to removal will risk damaging the building itself".[11] Typically, these plants are stranglers, whose roots can find ways through pores of the building to penetrate deep inside. Thus, it becomes difficult to remove the plants, even after killing them. Over hundreds of years, the original structure of the building has been displaced to a great degree, and thus removing these plants is considered problematic.

Admitting that there is no "one size fits all" approach, professionals advise a case-by-case approach to assessing risk of plant growth on buildings. This is reflected in the case of Hong Kong's wall trees, where the team of engineers inspected and evaluated each of the listed sites for their threat to human life. But at the same time, a report by English Heritage brings out that not all plants pose a threat to walls. From its six-fold classifications of spontaneous wall fauna, it states that while woody species and woody herbaceous plants almost always cause damage, other plants which fall under the categories of soft herbaceous, ephemerals, lower plants, and climbers do not pose serious risks as far as the structure is concerned. Particularly, it states in the case of lower plants (lichen, mosses, algae, liverworts):

> *These plants do not generally damage walls and should be treated in the same manner as annual and ephemeral plants, being left untouched unless there are clear and specific reasons for their removal.*
> *(Cathersides & Parker, 2014, p. 4)*

Sometimes, plants may even protect the building from climatic stress.[12] Ruined walls have large exposed pores which are like "open wounds" that make the structure prone to further dilapidation due to heat, rain, or snow.[13] In the UK, there is the practice of "soft capping", where ruined walls are protected from extreme weather by placing grassy turf on top of the walls (Cathersides & Parker, 2014). In addition to weather protection, the turf may also serve as micro-habitat for other plants and animals.

The ownership of the property is a factor, especially in case of public properties, as the government most certainly cannot take risk to public life. However, this is a grey area when it comes to private properties. Property owners who have affinity to trees growing on buildings said they would be tolerant to some degree of the plant growth since they find them aesthetic. These plants also attract insects and birds which can be considered a nuisance (and carriers of disease). However, people who like plants are likely to be okay with them. As one interviewee pointed out "all plants attract different insects and worms, sometimes really weird ones (amused). We are okay with all that. But there may be others who would like them less I suppose".[14]

In comparison, an assemblage of plants along the roadside or in parks are typically[15] not associated with major health risks. They may however cause allergies.

Decay and statuses of elements

Dereliction and decay are perhaps not created by plant growth or by dilapidation of a few buildings alone – rather, compounded by traces of everyday life around assemblages of biological growth. Located along the banks of the Ganges, Kumartuli in Kolkata is well known worldwide for its economy around traditional arts. Not all lanes of Kumartuli have buildings that are dilapidating. But a strong sense of decay emerges from the knowledge that with railroads and highways becoming more active as trade linkages of the city, thriving mercantile activities of the 1800s along banks of the Ganges have now largely disappeared, if not completely. Households have either out-migrated to parts of the city like Salt Lake which are constructed post-independence or have entirely left the city altogether – abandoning their properties. The conditions of poverty are reflected not in the elements of iron and cables or the exposed masonry – but in the observation that there are no traces of upkeep of the walls; no additional cost spent in planning during installation of the utility cables; and in the presence of iron casted into a trolley which is typically used by workers to carry goods across the blocks, that enable them to earn a minimum wage for their services.

However, visual appreciation is rooted in our subconscious beliefs of what is plausible for the locale. In cities of the UK, landscape managers prefer to maintain naturalistic landscapes for delivering higher levels of ecosystem services and because people of the area prefer the wild aesthetic. For instance, when I showed the picture of a very small cemetery in Oxford that had wildflowers growing within it, those familiar with the city's context were quick to question, "How can you be sure it is not managed?" Indeed, I wasn't sure. Landscape managers could have well intended to maintain it that way.

The species of plants involved in creation of an assemblage matter. Certain plants get statuses of weeds, which make them particularly undesirable in the eyes of designers and landscape managers.[16] For instance, wild flowers in Bengaluru have grown rampantly over the city colonising the city's public landscapes. To save resources the city has generally preferred low-intensity landscape management. Despite a preference for retaining naturalness, there is little tolerance for the weeds. Flowering plants exotic to Bengaluru, such as *Lantana Camara* while considered weeds on one hand, also draw pollinators such as butterflies and bees which feed on their flowers. While some regard that mosses add to the age value of buildings, a common practice has been to remove them.

Plants that may naturally grow on (heritage) walls are "specialised species"[17] but are not necessarily classed as natural heritage themselves. There is a slight possibility of these sites harbouring rare species, e.g., rare snails have been reported to be found living in walls of Dover Castle, in the UK (Cathersides & Parker, 2014). However, the rarity also means that it is not common for practitioners to come across situations where the protection of the biological growth is of an existential concern to species. To this end, one conservation architect based in India noted how trees that invade buildings are "*peepal* or sacred fig (*Ficus religiosa*), banyan (*Ficus macrocarpa*) and sometimes, *jamun* (*Syzygium cumini*)"[18] – which are abundant in their respective regions.

Managing growth in residual spaces on the ground is easier, cheaper, and less risky than compared to managing plant-growth on buildings. Managing them on buildings is not always worthwhile: 1) because they are very easily replicated with safer systems such as trellises and planters, 2) because buildings that are dilapidating in the cities are not always old enough to be considered heritage. While old buildings, especially in traditional areas of cities like Bengaluru and Kolkata, can easily be deemed historic, not all of them are classified as heritage. Unlike Whitby's castle which is maintained by the government for the purpose of the city's historicity, buildings in India unless declared heritage will not receive public sector assistance (via funds and other means) to maintain the sites.

"Dirty" environments

The poorly designed garbage dump in front of the plant-wall assemblage on Amherst Street creates a sense of disgust for visitors (Figure 7.5). Passers-by hurry past the spot in their best attempt to avoid the foul smell and the sight of a dirty environment, careful not to step on leachates that have been formed from leaking bins and residual waste. Some walk on the other side of the road, while others look away from the "dirty" area. Although I have highlighted before in Chapter 4 on Aesthetics how components that make up this particular assemblage are of rich art value, the garbage dump determines its perceived value. If the space was better designed, or allocated to another activity, the plant growth on the ornate walls could well become an iconic landmark.

Interestingly, the development of two sites is also entirely unrelated: the tree growth has happened in a private property with the partial oversight of the landowner, whereas the garbage dump is created by the public authorities where they could find space to do so. Small-scale nature growing in public spaces is also prone to littering – turning otherwise harmless green patches into sites of blight and decay. Thus, the role of regulating waste has a link with enhancing value of landscape fractions in cities.

FIGURE 7.5 Context depreciating art value: a roadside garbage collection point in front of a unique assemblage of wild plants and buildings at Amherst Street of Kolkata. July 14, 2018.

The context of Amherst Street is that of a lack of cleanliness, more than that of urban decay. Even though the buildings along the street from the colonial period have now started weathering down to allow plant growth, the large parcels of properties have high value in the eyes of real-estate developers. The neighbourhood is very near to the most renowned educational institutions of Kolkata and is well-connected to major shopping centres and public health facilities. If the properties were sold, they could become multi-storey housing or commercial complexes very soon. However, the owners are reluctant, as families have lived in the houses for generations. The massive buildings of the colonial period may now be difficult to upkeep, but still carry a reminiscence of power which the forefathers of these residents once possessed.

Hidden life of plants in cities

Plants have a hidden life in cities, as often botanical growth may be so small, that they continue to exist unnoticed; or they may occur in abandoned and restricted sites; or in spaces such as street corners and drains that fall outside the everyday eyesight of human beings.[19] Approaches normally used for framing nature in cities have not classified and mapped such assemblages of nature. Tools such as city master plans, green and open space plans, and tree-cover maps that document green cover in neighbourhoods drill down to the details of a pocket park, a roadside stretch, or even a single tree – they however do not account for microscopic nature. Furthermore, the temporality of such assemblages only makes it harder to discover and prepare systematic records of microsites. For instance, ephemeral growth may go unnoticed by those who did not happen to walk in that direction during the lifecycle of the plants. Then how do we frame landscape fractions in a manner that can induce some kind of consistent resource in planning and design? One might think that photographs are better for representation of reality and they can be deployed very quickly to gather data – as I have done in this book. Others may opt for "living maps" that crowd source information to offer a more flexible approach to documentation. But no one approach may be enough, and there is a need to use a combination of tools for the task.

Even though the existence of such assemblages is often microscopic, to discover and understand landscape fractions in cities, the approach to documentation would have to be defined with a set of scales that can adequately capture the relation of these sites with their immediate environment. This may become a complex exercise, for Schuster (2013) argues that while aerial shots remove connection with humans, they reflect the link with nature and the built environment. A bird's-eye view of the neighbourhood street side from atop a building, would indeed make it easier for one to perceive how small green islands are formed of wild growth amid large tracts of impervious surfaces (Figure 7.6). However, "top-down" views do not always capture adequately the lived realities of urban spaces. This kind of knowledge may lie in traces (e.g., playing cards or paper cups) which offer clues of how the space is used. The plastic and paper litters are to be cleaned up by municipal workers, who would have the privilege of experiencing, up-close, the different plant species that compose the green patches, or small organisms which momentarily visit these landscapes. Knowledge of wild spaces also lie in multiple temporal glimpses that help understand how residents, street vendors, herders, operators, mechanics, children, and animals use the same site in different times of the day, or how plants grow and dry out during different seasons.

FIGURE 7.6 Aerial view of the control box in Bengaluru, islanded by wild growth around it. June 20, 2020.

It becomes necessary to consider that it is not nature alone which is transitory – it is also the human beings who are only there to enjoy for a moment. Pedestrians don't notice everyday compositions of nature or those assemblages which are tucked away behind compound walls, control boxes, or even inside open drains (Figure 7.7). To gain access to such nature is to search for them, much like how one would look for rare species in complex natural settings such as forests or grasslands. In a city, however, a researcher with a camera can document such nature by peeking into drains and asking a landowner's permission to explore his building or his yard. Photographs can draw out the hidden microsites and present to the public that which otherwise are not visible to the naked eye. One can draw parallels about this to what Henri Lefebvre in his book *The Production of Space* refers to as "underground and repressed life" or "unconscious" of the city (Lefebvre, 1991).

If landscape fractions are not noticed, they are also less likely to be part of urban development discourses. Thus, the art of framing is inextricably not just linked to photography but also exhibitions that can

FIGURE 7.7 Nature growing inside corner of an open stormwater drain in Bengaluru. August 11, 2018.

draw out meanings and, in a sense, allow participation of nature in society. Photographs can be selected to recreate enjoyable experiences culminating from retaining a fleeting moment of green that would have disappeared later in reality. Such transient captures evoke the nostalgia of rain or magnify smaller insects and flowers to appeal to visitors in an exhibition gallery. Leet (2003) explains how nature is constructed through images and how emotionally charged photographs can facilitate political ends. One political goal of the environmentalist is to help people recognise everyday nature. The process of scientists and environmentalists to communicate knowledge of wild spaces in cities to the general masses can be pictured as a sort of awakening from the "unconscious"[20] state of the city.

Photographs, however, are not just for documentation, but are subject to interpolation that matches what the society desires to see (Niedenthal, 1993). They can be understood as a powerful medium of mass communication today – an academic captures the proliferation of images in our everyday life:

> *Landscape photography appears on or in screensaver backgrounds, TV commercials, cover art for apps or music, marketing campaigns, posters, textbooks, waiting rooms, corporate offices, hotel rooms, departmental promotional materials, study abroad brochures, among other materials in commercial, public, and private spaces.*
>
> *(Devadoss, 2022, p. 26)*

Through interpolation and social production of images, landscape fractions which are tucked away, or are only seasonal can become a more permanent part of everyday life. There has been a growing curiosity of the photographers and public with abandoned sites. Some examples would be ghosted fishing villages in China or decommissioned amusement parks in the USA. Images of these – often digitally enhanced – can easily be found on dedicated websites for curating photographs, or on social media accounts of enthusiasts. Such photographs may be a depiction of strange encounters, or of the uncommon aesthetic associated with wild urban spaces. In these captures of abandoned built-environment, plants can be seen to have grown over the structures of joyrides, buildings, and so on. But as examples from this book should illustrate, that spontaneous nature is not just found in isolated and restricted environments, and often has a greater proximity to social lives of human beings in cities – making them an integral part of everyday urban nature.

Mere photographic documentation of wild spaces will not by itself advance our understanding of these built-nature assemblages. Treib (2008) points to how much of photography is being produced for the sake of producing, yet "without necessarily effectively engaging the brain" (p. xix). Noticeably, at the moment, there is very little discourse on wild spaces in cities, let alone small-scale wild growth which are virtually invisible to the naked eye. In absence of this discourse, the functioning of these sites and their relationships with society has remained one of the biggest mysteries of urban nature. Over the course of this study, I found that news articles and popular media posts (e.g., blogs) about spontaneous natures by biologists and nature enthusiasts, almost without exception, limited their analyses to single strands of plants or species to distinguish weeds – missing out entirely on the context of space. Analysts and commentators of built-environments on the other hand used spontaneous nature as a means to indicate the poor state of real estate affairs. The dominant discourses seem to be revolving around their demerits, even when the context of discussion is outside risks to human life.

The social lives of wild spaces remain undiscussed. When they are discussed, and if at all, the same narratives of decay and risks are repeated. There is both redundancy and reductionism in our understanding about these natures which is narrowing the space for constructive dialogue. To help frame the potential of these fractions in a more balanced manner, there needs to be more creative discussion around these assemblages. Such discussions may be supported not just through research in the different scientific disciplines, but also through the imagination of artists, poets, and writers of today who can challenge the general social conscious. Exhibitions and displays of magnified photographs of small-scale spontaneous nature can enable people to recognise and construct discourses around nature which they otherwise overlook. Matthew Gandy in his book *The Fabric of Space* discusses in detail how photographic documentation of Paris's sewers by Félix Nadar influenced society to view the otherwise invisible, yet critical, infrastructures of everyday life and begin appreciating them as engineering marvels (Gandy, 2014). There is no doubt a potential for the same with spontaneous nature, as exhibitions could certainly add force to their bringing forward their variegated and fragmented existence through collections of landscape fractions under one roof – particularly those that grow tucked away on street corners or hidden away from eyesight inside storm water drains.

Others have argued how photographs are deliberate constructs rather than accidents (see Treib, 2008). Indeed, clicking a picture to share as a social-media post can be a very spontaneous response. But many of the interesting assemblages can occur in old town areas or in heritage sites, where people might take their pictures as part of a more organised troupe. Perhaps stories associated with these sites can make them more interesting and valuable additions to citizen-led heritage and urban nature walks. These semi-guided tours are great tools to bridge the gap between research and practice. They are also great tools for sensitization of the local communities. Virtually just about anybody can sign up for these tours, and explore the historicity, aesthetics, and meanings of landscape fractions.

Notes

1. Snodgrass and McIntyre in Thomas Christopher's edited book short-titled *The New American Landscape*, records how "during dry spells, hardy succulents such as *Sedum* species suspend daytime evapotranspiration, effectively conserving water in their tissue, while most herbaceous plants, attempting to minimize surface exposure, wilt and soon die. Even plants that go dormant rather than dying completely, such as some grasses, take longer to regrow, and they produce a lot of dry biomass, which in addition to being unattractive can be a fire hazard in some areas" (Snodgrass & McIntyre, 2011, p. 167).
2. In her article, *Guidelines for Tree Photography*, Heather Angel writes: "The photography of whole trees in their landscape should present few problems, since the subjects are not only comparatively large, but they are also static. Plenty of time can therefore be spent selecting specimen worthy of photography" (Angel, 1980,p. 105). The same can hold true for architectural subjects which are also immobile.
3. See Chapter 2 on Existence for identities of assemblages.
4. See Chapter 4 on Aesthetics for more discussion on this example.
5. Erving Goffman's book *Frame Analysis* offers a sociological interpretation that lays the foundation for studying roles of context in production of discourses (Goffman, 1974).
6. Michael Busselle and John Freeman in their co-authored *Photographer's Handbook* discuss "isolating the subject" as a strategy (Busselle & Freeman, 1984, p. 106).
7. Abdel-Raheem (2017) holds that images are a "intricate web of connected frames" (p. 333), and suggests that meanings of pictures are constructed through influences from wider discourses in the society.
8. This term is generally used for two-dimensional art. I relate it to the building façades.
9. Perhaps a useful reference for those wishing to explore landscape composition – British watercolour landscape artist, Leonard Richmond, in his book

Essentials of Landscape Composition demonstrates composing landscapes with clouds (Richmond, 2012).
10 Cf. Davanzatti, Leçon sur la monnaie (Quoted in Foucault, 2005, p. 188).
11 P. Das. Heritage conservation and plant growth on buildings, interview with A. Deb, December 3, 2021.
12 Writing for English Heritage, Cathersides and Parker (2014) note how ivy plants protect structures from rain and frost.
13 H. Vangara. Effects of plant growth on buildings, interview with A. Deb, January 11, 2022.
14 P. Das. Heritage conservation and plant growth on buildings, interview with A. Deb, December 3, 2021.
15 J. Munirathinam. Ecology of small-scale wild growths in cities, interview with A. Deb, November 23, 2021.
16 K. Keswani. Appropriation of small green spaces, interview with A. Deb, December 4, 2021.
17 Writing for English Heritage, Cathersides and Parker (2014) note "True wall plants are able to withstand extremes of temperature and drought" (p. 2).
18 P. Das. Heritage conservation and plant growth on buildings, interview with A. Deb, December 3, 2021.
19 See section on invisibility of everyday nature in Chapter 5 on Collectives.
20 Noting the potential of this awakening within a city, Henri Lefebvre writes "If it turned out, for instance, that every society, and particularly (for our purposes) the city, had an underground and repressed life, and hence an 'unconscious' of its own, there can be no doubt that interest in psychoanalysis, at present on the decline, would get a new lease on life" (Lefebvre, 1991, p. 36).

References

Abdel-Raheem, A. (2017). Decoding images: toward a theory of pictorial framing. *Discourse & Society*, 28(4), 327–352. doi:https://doi.org/10.1177/0957926517702978.

Angel, H. (1980). Guidelines for tree photography. *The International Journal of Urban Forestry*, 4(2), 105–114. doi:https://doi.org/10.1080/03071375.1980.10590599.

Busselle, M., & Freeman, J. (1984). *Photographer's Handbook*. London: Octopus Books.

Cathersides, A., & Parker, E. (2014). *Landscape Advice Note: Vegetation on Walls*. English Heritage. Retrieved January 3, 2023, from https://www.english-heritage.org.uk/siteassets/home/learn/conservation/gardens-and-landscapes/lan_-_vegetation_on_walls.pdf.

Davern, M., Farrar, A., Kendal, D., & Gunn, L. (2017, May 4). *Higher-density cities need greening to stay healthy and liveable*. Retrieved January 3, 2023, from The Conversation: https://theconversation.com/higher-density-cities-need-greening-to-stay-healthy-and-liveable-75840.

Dee, C. (2001). *Form and Fabric in Landscape Architecture: A Visual Introduction*. New York: SPON Press.

Devadoss, C. (2022). Monolith, the colonial face of landscape photography. *Gender, Place and Culture*, 29(1), 26–51. doi:https://doi.org/10.1080/0966369X.2020.1842334.

Foucault, M. (2005). *Order of Things* (Reprint ed.). New York: Routledge.

Gandy, M. (2014). The Paris sewers and the rationalization of urban space. In *The Fabric of Space: Water Modernity, and the Urban Imagination* (pp. 27–54). Cambridge, MA: MIT Press.

Gandy, M. (2016). Unintentional landscapes. *Landscape Research*, 41(4), 433–440. doi:10.1080/01426397.2016.1156069.

Goffman, E. (1974). *Frame Analysis: An Essay on the Organization of Experience*. Boston: Northeastern University Press.

Leet, S.-K. (2003). Nature as construct: a conservation project in Singapore. *Photography and Culture*, 6(2), 235–243.

Lefebvre, H. (1991). *The Production of Space* (3rd ed.). (D. Nicholson-Smith, Trans.) Cambridge, MA: Basil Blackwell.

Niedenthal, S. (1993). "Glamourized houses": neutra, photography, and the Kaufmann House. *Journal of Architectural Education*, 47(2), 101–112. doi:https://doi.org/10.1080/10464883.1993.10734582.

Richmond, L. (2012). *Essentials of Landscape Composition* (This Dover edition first published in 2008, is an unabridged republication of *Essentials of Pictorial Design*, originally published by Sir Isaac Pitman & Sons, Ltd., London, in 1933. ed.). New York: Dover Publications.

Schuster, J. (2013). Between manufacturing and landscapes: Edward Burtynsky and the photography of ecology. *Photography and Culture*, 6(2), 193–212. doi:10.2752/175145213X13606838923318.

Snodgrass, E. C., & McIntyre, L. (2011). Green roofs in sustainable residential landscape. In T. Christopher (Ed.), *The New American Landscape: Leading Voices on the Future of Sustainable Gardening*. Portland, OR: Timber Press.

Treib, M. (Ed.). (2008). *Representing Landscape Architecture*. New York: Taylor & Francis.

8

IMAGINARIES

Competing visions of the urban future

The New Urban Agenda[1] of United Nations has laid foundation for sustainable development of cities. Among other sustainable development goals or SDGs, cities should ensure health and wellbeing (SDG 3), climate action (SDG 13) and life on land as well as under water (SDGs 14 and 15). To achieve this, architects and planners are thinking of enhancing green infrastructure in cities. The visions or imaginaries often feature homes with roof gardens and backyards, lush green street sides around the homes, and well-maintained parks and playgrounds in the neighbourhoods. These visions of green and picturesque urban environments are well accepted globally, and it is also well agreed upon that cities must accommodate more nature to become more sustainable.[2] Although a variety of practical considerations limit the actual implementation of these plans that strive to achieve the imaginary of a "sustainable city", the costs required to maintain the landscapes, or the aesthetic preferences of people who opt for dry landscaping with very little or no green would be two such limitations. Transformations of green to grey may be observed in large town squares which are almost entirely paved, with sparse growth of grass and perhaps accompanied by some potted plants; or neighbourhoods where residents have paved the green lawns of their front yards to ease maintenance; or where the

DOI: 10.4324/9781003374602-8

development authority has poured concrete over roadside verges to maximise space for automobiles. The spatial patterns that have thus emerged signal a challenge to provisioning access to greenspaces within cities, and in their current form, hinder the goal of making urban communities sustainable (SDG 11).

The globalisation of urban aesthetic may be noticed in international business districts, shopping malls, and high-end residential complexes – it comes with perceived cleanliness of say, for example, having concrete paving or highly manicured ornamental landscaping as opposed to green spaces, which are muddy and ridden with rotting leaf litter. The proliferation and universal acceptance of such aesthetic as city goals has made it difficult to challenge them, no matter how unrealistic for certain communities. Left unfulfilled, the grand vision of a "beautiful city" facilitated by global aesthetics of landscape design, may be contrasted by another vision emerging out of people's needs, that looks at utilities of green spaces differently. This alternative vision may be thought of as a grassroots appropriation of green spaces; it may be characterised by grazing, foraging, and other uses, otherwise excluded from spaces with global aesthetics (discussed in Chapter 5 on Collectives). The grassroots vision can be understood as a more pragmatic reflection of the city's fabric.[3] Say, for instance, the experiences from Bengaluru discussed in Chapter 5 suggest that there are still communities that continue to depend on agrarian activities for livelihoods within urban areas. In doing so, they appropriate neglected spaces of public green infrastructure – such as parks and roadsides – for their own use. Given the diminishing quantity of green spaces as well as rigid management regimes of urban greens, activities such as grazing and foraging can only be practiced in small pockets within the city. These activities are often dependent on informal urban vegetation. Ignoring such critical links and wiping out small-scale assemblages of spontaneous nature would mean destroying valuable life support systems of the poor and low-skilled population who are particularly vulnerable to rapid globalisation of cities. Instead, assemblages may be better managed to help create ladders of growth.

Another example of grassroots vision is how spontaneous nature becomes a mode of greening in certain neighbourhoods that are otherwise struggling to afford the upkeep of green spaces in accordance with the global aesthetics. This is because public funds that help maintain landscaping are limited, and are often directed to well-performing districts as a commitment to equitable remittance of revenues back to ensure districts continue to keep performing. One fallacy to this approach is that even with neoliberal freedom of capital flows, not all parts of the city will have a growing economy. Noticeably, parts of cities undergo rapid transformation attracting a vibrant global economy, but pockets within cities also exist that take a long time to change – these are slow growth areas.[4] This means while some neighbourhoods are going to have more well-maintained green spaces than others, some might not have them at all. To some extent then, this brings into question redistributive justice of the government and its obligation to deliver public goods to the rest of its citizens.

At this rate, cities are not entirely sustainable. They are becoming greener, but not equitably. Consequently, people living in areas that are devoid of everyday nature do not enjoy health and wellbeing benefits. Moreover, grassroots efforts of greening are thwarted by global visions. Indeed, the park which I studied to be a grazing ground for sheep, was later revegetated using ornamental plants and subsequently strictly monitored to prevent entries of cattle. Not only has the wild aesthetic disappeared, but a giant resource for local herders is now lost. Perhaps as Shackleton, et al. (2017) highlight, having political space for people to share their needs can help decision-makers understand how spaces are used. Consequently, compatible activities can be allocated for public land resources, without disrupting the social life of these spaces.

Even as the responsibility of service delivery is rapidly shifting to be market-driven, it is unrealistic to assume that privatisation of the housing sector will be able to take care of the problem of low green spaces. For sure, some of the burden can be shared by the developers and resident associations by provisioning and maintenance of green spaces. But these enhancements of the built-environments come at a cost. The costs of designed landscaping are finally borne by residents, as developers pass the costs on to the buyer. It must be considered that some buyers may not be able to afford these increased prices, and some buyers may not be willing to pay for green amenities. Particularly in the Global South, there are challenges of controlling unplanned developments and violation of building regulations that lead to over-building and encroachment into what should otherwise be green spaces. To some extent these violations reflect the demand for space that cities have been unable to supply, and consequently they may be regularised by the authorities. But in the process, green spaces which are equally important are deprioritised – leading to a rather obvious disconnect between people and nature. As a result, not all neighbourhoods are green and, particularly, the low-income populations struggle. Under these circumstances, spontaneous nature in general becomes a mode of greening (see Chapter 2 on Meanings). In places, where humans have been unable to maintain green spaces, the forces of Gaia have been able to revegetate the urban surfaces creating natural fractions that offer ecosystem services to help achieve the SDGs.

In the age of globalisation, designed versions of the "green city" have delegitimised assemblages of spontaneous nature as a mode of greening. Moreover, city dwellers have become familiar with small-scale assemblages of spontaneous nature as "transitory", "ugly", "risky", "wastelands", and so on. These broad and arguably unchallenged generalisations have led to the exclusion of spontaneous nature from sustainable development goals. This notwithstanding a more recent body of work – including discussed concepts such as ambivalent and unintentional nature of landscapes – has emerged, pointing to a better understanding of the role of wild spaces in urban development. The framework of landscape fractions expands upon these to systematically explain small-scale assemblages specifically. The framework should encourage people to view these natures differently. In the following sections of this chapter, I will recap alternative imaginaries that run through different chapters in this book, to offer a more systematic and comprehensive understanding of small-scale spontaneous nature in cities.

Ambivalence of small-scale nature

Ironically, aesthetic assemblages of spontaneous nature in cities occur because of a certain absence of human oversight. Often these are in residual spaces within populated neighbourhoods, or neglected spaces such as parks and old buildings. Other times they may be found in abandoned spaces such as industrial yards and buildings. These conditions point towards dereliction and decay, where the value of an aesthetic assemblage can stay hidden. Such spaces otherwise may be considered wastelands and at the same time holding values of heritage, habitat, and potentials for making small urban places and community identity. This simultaneous existence of duality of good and bad is what Jorgensen and Tylecote (2007) describe as ambivalence of wild landscapes. There are certainly undesirable effects of having plant growth on buildings – they are capable of creating dangerous environments for habitation; and some residents may also not enjoy the insects, birds, or animals they draw.[5] However, those who have a greater affinity to nature might also be more tolerant to insects and bugs. Others may view a plant growing on a wall as a "green umbrella" of shade for pedestrians walking along it.[6] This seems to be a cycle, where more exposure to nature can create affinity to nature, and more affinity to nature would mean having more nature in cities.

It is common for assemblages to carry remains of influential architectural legacies, e.g., from post-industrial and colonial eras. However, the disappearance of such structures with high heritage value is common. While a movement has begun with the aim of preserving and reusing sites of industrial accomplishments, landscape fractions are not on par due to several limiting differences. The industrial monuments are colossal in size and are in most cases in close proximity to each other, because industrial facilities have typically been separated from residential and other areas. Thus, we notice them very easily, or can't miss taking note when they disappear.

Small-scale assemblages of nature are at most times not noticed because they are also fragmented, seasonal, growing in unexpected pockets across an entire city. The "hidden life" also makes it less obvious to take notice when fractions go missing. The disappearance of landscape fractions is discounted from the assessment during reconstructing any site because they are also considered transient and out of place – at once rejecting any possibility of future place-making. It is as though it is nature's responsibility alone to find, advocate, and make a place for itself. But the broad case of such assemblages is, nevertheless, similar to the human-built technological and architectural legacies which are also in need of recognition and need rethinking in terms of how to be reused and preserved as opposed to being dismantled to make way for new developments (see Loures, Burley, & Panagopoulos, 2011).

The problem, perhaps, with ephemeral growth is if they are relevant enough to be mapped as biotopes. Several approaches can be taken to map them:

a Delineating broad zones where these natures can be found more often. E.g., areas of low fiscal transfers, old fabric of cities, obdurate industrial areas, and so on. This book highlights in detail several potential areas where one can find assemblages of spontaneous nature, and the forces behind their existence.
b Crowd sourcing information to keep a more micro-level "living map" of such natures. E.g., using applications on mobile phones

where people can add, rate, edit, and update the locations of assemblages using photographs. Early versions of such applications already exist for people to help locate mushrooms or berries.

The book offers a framework for the viewer to judge if an assemblage has enough value to be preserved as a "landscape fraction". This does not necessarily imply that the approach is entirely anthropocentric. For instance, valuing assemblages for their potential of serving as ecological habitats may be more eco-centric. Notwithstanding, biologists, nature enthusiasts, and foragers would likely hold some of these fractions to higher value than compared to others who are not directly linked to such nature.

It follows naturally that it is possible some of us may even overestimate the value of small-scale assemblages of nature. For instance, the land may be put to better, more economically productive, uses, i.e., reconstructing buildings may allow new uses which offer more profits to the owner; or wild spaces may be replaced with more thoughtfully engineered landscapes that perform better in terms of climate regulation or aesthetic appeal. Having well designed small-scale green spaces may also prevent unnecessary stagnation of water and avoid plants that cause allergies or harbour insects that are irritants or are disease causing. Thus, it is quite clear that more research is needed to understand and distinguish values of small-scale assemblages of spontaneous nature from their harmful effects.[7] More importantly, knowledge has to reach those citizens who derive more benefit or are affected every day by the nature around them.

Shifting to a more optimistic view, spontaneous landscapes have many values. The ambivalence framework allows one to ask: what is the best use? First of all, there can be multiple uses of the same site. The sites may be of educational value, scientific value, aesthetic value, other cultural value, or simply for preserving autonomous animal and plant habitats. However, some activities may not be compatible with each other: say, for instance, grazing and aesthetics or autonomy of non-human life versus foraging. Some activities may also not be advisable for human interaction: sites used for education can conflict with risks of getting bitten by a snake, stung by scorpions, and so on depending on the geography. On the other hand, a typical risk may prove to be of lesser concern than perceived, e.g., tree growth on a building may be more of a habitat than a risk, if no one resides in the building. They all point to a need for a case-by-case analysis of such assemblages.

Ambivalence makes it hard to govern wild spaces, as people may be thrown into a dilemma to find the right course of action for a given patch of land. Furthermore, even though assemblages affect communities, dealing with them can become complicated because of varying patterns of land ownership. For instance, tackling wild growth in an abandoned private property posing risks of harbouring pathogens may rest with individual landowner(s); or managing wild growth along the roadside verges, may rest with the government rather than the residents. This does point to the need for stakeholders interacting and sharing concerns as well as ideas with each other. On one hand, such exchange has potentials of building community cohesion and on the other, spontaneous urban nature can be managed in a more democratic[8] manner.

Sites of experimentation and discovery

Landscape fractions that are small scale are easily found in everyday settings. These sites challenge the usual and the ordinary to open up imagination and possibilities. Gandy (2016) highlights how marginal spaces and abandoned sites can become sites of "adventure, creativity and discovery" (p. 433). Such potentials of abandoned industrial sites and spontaneous nature have been discussed separately before (Edensor, 2005; Jorgensen & Tylecote, 2007), but rarely have people discussed how industrial elements abandoned in residual spaces along street sides and other wild spaces, have created aesthetic assemblages worthy of similar values. Some examples which I note in this book are fallen tree branches that become benches for sitting and abandoned artifacts of mechanical engineering fuses with nature to create assemblages that leads the mind to question the logics of order.

Plant growth on walls is generally perceived to be a risk to the integrity of built structures. As in previous chapters, I explained how this is not always true. Moreover, while biological potentials of plants growing on vertical urban surfaces has been somewhat researched,[9] their social potentials largely remain unexplored. So now, I point to at least two alternative positive outcomes emerging from risks:

A. The risks of spontaneous nature have facilitated entrepreneurship and innovation. In some instances, the walls and other sites of old city and decaying fabrics have become laboratories for urban experimentation: conservationists frequently have to engage with these sites for removing plant growth on buildings to protect structures. However, instead of killing the trees, some choose to rescue and replant them. Such rescue operations may not always be successful or feasible, yet these sites have acted as grounds for experimentation to give rise to a valuable enterprise. In the city of Mumbai in India, a few citizens who gained experience in rescuing trees through such experiments, ended up registering an organisation for the cause and have reportedly handled over 3000 cases over a period of ten years (Bhalerao, 2020). That also means at least 3000 plant-built assemblages had existed in the city.

B.Risks of spontaneous nature have also brought out human intentions of co-existence and respect for non-human life. In contrast to rescue operations which tend to return plants to their traditional habitats, sometimes, plants growing on buildings had to be managed without removing in order to save the building. This notwithstanding, by managing biological growth the risks exerted by growing plants on the built structure, don't entirely go away. However, Hong Kong's example explained in Chapter 3 on Existence suggests one option of dealing with this risk as monitoring by competent engineers. Perhaps more research can help deduce early warning signs to help mitigate risks. Furthermore, despite risks, people have found ways to appropriate such settings for cultural purposes; they have inspired people to film movies and undertake haunted walks.

Unusual objects can be found along with spontaneous growths that make such sites interesting. Within a residential neighbourhood somewhere in outskirts of Bengaluru, lay an abandoned model of a helicopter (Figure 8.1). The blue shed in its background belonged to an aviation school located right opposite the street. Returning to the illogic of space, it is strange to find the body of a helicopter, on the street side of a residential neighbourhood, rather than on an airfield. The rarity of the occurrence may inspire poetic thoughts about the grounded flying machine without wings to reflect an "object's desire to fly, but no hope of doing so".[10] Nature is an integral part of this analogy to the observer as it establishes the notion that the site is abandoned or outside human intervention. The mere wild growth of grass has an overbearing on setting as consequently, it was identified that "if [the helicopter model] was instead placed on a factory floor then [it would signify] hope positively, because it gives an impression, the wings will get fixed eventually".[11]

FIGURE 8.1 Machines as part of post-industrial landscapes: a model-helicopter kept on a street side verge in Bengaluru, blends with wild plants underneath, and around it. September 27, 2018.

Some artists and conservation architects are exploring old structures which are considered ugly or are yet to be included in the heritage preservation conservation. A New York–based artist, Otero-Pailos, points out that the word "experiment" carries the risk of failure – and that failures in conservation carry high costs both historically and culturally (Otero-Pailos, 2016). But as opposed to fragile artefacts such as paintings, the extent of risk on experimenting with architecture of buildings, and plants growing on them, are somewhat mixed. As I have highlighted before through examples, some buildings with plant growth are hazardous, yet inseparable. It is perhaps a pertinent reminder that in such cases, preservation of structures has typically been done while retaining and managing plant growth.

The response to revitalising post-industrial landscapes which include designing green parks and public spaces around pre-existing built elements which become monuments of the industrial era, draws people to appreciate their historicity and engineering of the otherwise derelict structures. Wu (2011) from China argues the need for integration of cultural and economic productivity with these sites of industrial heritage. Studying North American countries, De Sousa (2014) observes that although public, private, and non-profits are increasingly taking interest in greening post-industrial landscapes, they are faced with challenges of land acquisition, sustainability of funds, and maintenance, as well as reducing risks that arise from a derelict space (e.g., soil contamination).

Other surprises await people who seek to explore small-scale wild spaces in cities. Microsites may bear endangered species to the interest of scientific communities (see section on mapping biotopes of urban nature in Chapter 1 on Morphology), or insect kingdoms to appeal to enthusiasts (see plates on red cotton bugs in Chapter 5 on Collectives), or pretty flowers (see section on unintentional landscapes in Chapter 4 on Aesthetics) to please the eyes of a general passer-by. These make the small-scale spaces special in the sense that they are able to reconnect the fading relationship between nature and human beings in cities.

In the absence of obvious resources and framework, how some citizens have incorporated these spaces was refreshing to find during the fieldwork. People have managed to add art value by graffiti, yard design, etc. The risks have been mitigated through periodic monitoring and management of growth. Other times the process of co-existence is just accidental – e.g., flowers around the control box. This is why I distinguish the concept of co-production from co-existence in Chapter 6 – with the human appropriation of wild spaces in co-production being the key element of distinction.

Reconstruction of meanings

Assemblages of spontaneous nature and built environment in cities are often framed through dominant discourses that surround elements in the assemblage along with its spatial and political context. Positive meanings of spontaneous nature have been less discussed and supressed due to the overbearing of the significance of their negative externalities. It is not uncommon for spontaneous growth in cities to be related to economic success and decline of cities, as wastelands, as grounds for invasive species (Del Tredici, 2010). However, the set of views also seem to have developed from a scale that considers the city as the object that is situated within a system of other cities across the globe. Due to this top-down perspective, the meanings of small-scale spontaneous nature have often been compared with more prominent modes of greening such as designed landscapes or gigantic natural settings. This has led to a skewed understanding of landscape fractions that measures these unique assemblages through discussing what they are not – instead of taking into account what they are, or even what they can be.

Because of their small scale and fragmented existence, landscape fractions may not contribute substantially to the aesthetics of a place, habitat, climate regulation, or economy of the city. But this does not discount the fact that they continue to be a support system offering stepping stones and corridors for species movement between larger urban ecosystems; they play a vital role in greening urban spaces which offers health benefits no matter how small a role in climate regulation; they are not just composed of small plants but at times, also large trees that contribute to climate regulation; they also offer ancillary incomes and food from livestock and edible plants; and they may also occasionally contribute to both aesthetic and utilitarian placemaking in cities. Thus, a bottom-up approach to studying these landscapes would allow us to better evaluate their benefits.

There exist several piecemeal practices, that show the way towards engaging more meaningfully with small-scale sites of wild nature. Scientists have used them as "Island Laboratories"[12] or grounds for research. Children explore them to quench their curiosity about nature by observing animals and plants (sometimes consciously, and other times unconsciously).[13] Yet urban planners and designers are yet to engage in these sites and incorporate them into city building processes.[14] While suggesting concrete policies is beyond the scope of this book as they depend on the context of each city, by merely studying spontaneous nature I feel these assemblages hold valuable knowledge that would help urbanists understand cities better. Whether as aesthetic sculptures of plant-built ensembles, or as indicators of socio-economic development, wild spaces in cities offer too diverse a potential to be left out of a city's plan preparation processes.

On the prima facie, the assemblages are islands, i.e., they are fragmented and spread across the city like tiny dots. However, Chapter 5 on Collectives brings out the connectors between them as flows of air, water, nutrients, non-human life, and so on. Connectors help imagine these assemblages as collective ensembles that compound greater benefits at various scales. The idea of scales is important as visual assemblages can happen at the street level – not a resolution higher. For instance, the contribution of wild spaces to climate regulation may only be significant when aggregated at the city scale; but it is unlikely that a group of microsites in a neighbourhood can adequately add enough evapo-transpiration to cool the atmosphere. Some imaginaries of assemblages crosscut multiple scales and even fuse small-scale wild spaces with larger parcels of green infrastructure in cities: think of the grazing networks spread across multiple neighbourhoods that comprise both neglected parks as well as roadside nature; or think of landscape fractions serving not just as habitats but also as stepping stones to facilitate movement of insects, birds, and other small animals between green and open spaces within the city.

The meanings of spontaneous nature remain open to interpretation and largely unexplored. I offer examples of landscape fractions, not always from sites that are abandoned and locked out of bounds to citizens, but also from those sites which are part of their everyday lives – along the streets, around control boxes, on their walls and roofs, and so on. Sometimes, nature is visible through grills and other vantage points to help become part of people's everyday experiences. Other times, nature is tucked into niches and corners which are not restricted, but are out of regular sight. Often people get used to their everyday nature and thus are less likely to consider them noteworthy. Nevertheless, such nature help bring benefits of green (e.g., reducing stress and serving as a visual relief) to otherwise deprived neighbourhoods.

Negative implications of human-nature assemblages have, in several instances, been altered through simple interventions of managing biological growth, cleaning litter, visual uplift of built elements, and design interventions that improve the utility of spontaneous growth in cities (e.g., pathway amid wild grass in backyard). While the perceived spaces in cities reflected in planning and design guidelines have largely offered a safe distance of trees and buildings, many neighbourhoods that have existed before these norms are struggling to allocate space. At the same time, certain experiments show how people have appropriated small-scale nature along narrow spaces to offer green oases amid a densely built urban fabric, otherwise challenged to afford planting large strips of nature. Small quantities of nature can be viewed as a tool to retrofit otherwise deprived neighbourhoods in order to enjoy the benefits of green environments. To this end, spontaneous growth fused with buildings and technological relics, or growing in residual and neglected spaces are opportunities for placemaking and delivery of green infrastructure.

Of the positive meanings of landscape fractions, I explore several alternative narratives around these assemblages that contrast from their dominant imaginaries: from death and decay to new leases of life; from isolated and disconnected infrastructures to symbiotic and collective organisms; from unwanted growth to a mode of greening (see Chapter 2 on Meanings). The generalisations of risks have in many cases overshadowed benefits these sites offer, and consequently discouraged explorations that can discuss any productive use of these sites. Reconstructing positive meanings around these assemblages can help make balanced decisions of whether to retain these landscapes or do away with them.

Shifting status of wild growth

The statuses of wild growth flow from generally held meanings in the society. One would think that given assemblages of my interest have components of architecture and plants, scientists and architects would value them differently from the general public (see Angel, 1980). But my discussions with practitioners reveal that such is not entirely the case. In their professional capacity, conservation architects, landscape architects, architects, urban designers, and city planners as well as biologists are wary of suggesting positive contributions of wild landscape in cities. However, the same subjects in their personal capacity are appreciative of some of the assemblages which I showcased. The hesitation to associate with spontaneous nature points to the need for a shift of status of these arrangements within the city. By shift of status, I do not mean a blind appreciation for these landscapes. Instead, spontaneous nature is to be given a more complex evaluation for both their benefits and risks.

It is critical to understand the array of ecosystem services these assemblages offer, because it can help planners and designers think about how services of wild spaces can be integrated to benefit local communities. If statuses of assemblages are dismissed as useless landscapes, then potential sites may not be considered during planning processes, and it is an opportunity lost for the city. Assemblages are not devoid of their immediate physical context and wider discourses that can be associated with them. This is discussed in Chapter 7 on Frames. Current imaginaries surrounding spontaneous nature are largely negative – thus, a conscious effort is needed for a shift in mindset. In other words, there needs to be a shift of statuses of wild spaces from negative to positive: from places of decay to mode of greening; from spaces of neglect to spaces of multi-species co-existence; and so on. Landscape fractions act as a framework to analyse these small-scale assemblages of spontaneous nature in further depth – an analysis that has been lacking to date.

Small-scale spontaneous nature is not rare per se – they can be found almost everywhere. But it is safe to assume not all assemblages will be deemed as aesthetic and valuable. Unattractive wild spaces are often described as messy or untidy. Indeed, aesthetics can become a matter of individual preference. But in the case of planning for public good, one needs to understand the shared values of these landscapes. For instance, are some of these assemblages iconic enough to act as landmarks or do they bear enough historicity to be deemed as monuments? Others have argued for more utilitarian values along with the visual aesthetics. To understand these, values would have to be documented considering temporality (e.g., seasons of unattractive state) and across the ecological, cultural, and economic services these settings have to offer. But more importantly, there needs to be an awakening of sorts among people to recognise these values and begin accepting these natures as part of the city. Sometimes it is a matter of art or photographic exhibitions[15] that, through documentation and even interpolation, bring out details otherwise invisible to the naked eye; or eventisation of wild spaces through nature walks and incorporating them with heritage walks that help people discover interesting settings; or attention to discourses that (re) frame issues around the subject through media such as blogs, films, and literary works; thereby strengthening the place of such natures in our hearts and develop more rigorous and inclusive imaginations around these settings.

Wild growth in cities is not transitory – they are here to stay. There will be plenty of opportunities for small-scale assemblages of spontaneous nature to emerge and thrive as all buildings naturally degenerate, as developers prefer greenfield over brownfield, and as local governments continue to face austerity cuts for redevelopment and landscape management. The study of assemblages in this book, that has been conducted over multiple periods of time, affirms that spontaneous nature is not always transient nature – some sites survive much longer than others; this notwithstanding the seasonal variation of their aesthetics. Observations revealed that several sites have been withstanding anthropogenic pressure of transformation for multiple years, and are unlikely to disappear soon. Architects, urban designers, and planners should look at how these precincts could be integrated as sites of education and exploration – especially for raising the next generation of citizens who, given the lack of access to green spaces, otherwise might have fewer opportunities for reconciling with nature.

Unplanned biological growth is considered as informal. However, in many cases wild spaces were managed in an organised manner. Sometimes landowners and communities have outsourced paid landscape management services. Other times, government has recognised growth on buildings as integral to conservation of architectural heritage; and to ensure safety, these assemblages have been monitored by public appointed engineers. Therefore, landscapes can be co-produced through a mix of formal and informal management processes. In Chapter 5 on Collectives, I illustrated how designed and non-designed nature get mixed to create aesthetic assemblages. Typically, landscape designers take into account the seasonal variations of plant growth, and plan for them in advance, so that landscapes appear aesthetic year-round.[16] Such expertise may be used to complement spontaneous nature with design interventions that make up for the lack of aesthetics in off seasons. Landscape fractions are grounds where formal and informal boundaries of landscape management are blurred; and a growing thickness emerges in the relationship between humans and spontaneous urban nature.

Plant growths on built surfaces make novel compositions and are, therefore, not "invasive sites" per se. Yet they can become co-produced assemblages which capture values of memories and heritage. Partial remains of historic monuments can sometimes be found splintered into small fragments across the city. Neglected and hidden, these sites are often found with spontaneous growth fused with the structures. Combinations of plant growth on heritage structures can together form interesting sites that build on each other's values. As explained in earlier chapters, synergies may include age value, adding dynamism and life to otherwise static and dead relics, and so on. Sometimes the plant growth may be so deeply fused with built structures, they are hard to separate. In such cases, people have often chosen to retain the plant growth, so as to save the relics with historical and memory values. These sites exemplify as what I call co-produced.

To conclude, an overall shift in statuses of spontaneous nature is necessary to begin viewing these landscapes as an important and integral part of urbanisation. Yes, spontaneous nature points to rather obvious deficiencies of cities. For instance, neighbourhoods with spontaneous nature are certainly struggling with resources such as space and finance. But it is also true that greening has happened through non-human interventions and continue to offer benefits which help cities reach sustainability goals such as health and wellbeing, biodiversity, and so on. Temporarily left outside of human control, nature has taken over urban landscapes to restore a balance that was compromised when we replaced nature with concrete surfaces. This restorative force, better known as Lovelock's Gaia, can be a bane or boon to cities. On one hand, if cities ignore them, risks such as allergies or dilapidation of built environments may occur. On the other, if cities discard them or replace them, an opportunity to harness their potential is lost.

Notes

1 The New Urban Agenda was launched at the Habitat III conference held in Quito, Ecuador in 2016.
2 Green spaces are connected to people's health, offering animal habitats as well as regulating climate of the urban environment.
3 The term "grassroots view" has been described by Williams, Meth, and Willis (2014) as a category of development ideas outside of the typical state-led and market-led line of discourses. In doing so, the authors use the term "autonomous development" to suggest the role of communities (along with civil society and NGOs) as the key actors in defining development, in absence of, or in partnership with, the market and state actors.
4 Considering growth at the city level, Leo and Brown (2000) point out that urban development approaches tend to impose policies of rapidly growing areas onto slow areas, whereas a different set of policies may be more beneficial. The complexity in adopting such alternative policy approaches onto decision making and quality of life in slow growing cities has revealed itself over the past two decades (Mike Raco, 2018; Hayes, 2006).
5 P. Das. Heritage conservation and plant growth on buildings, interview with A. Deb, December 3, 2021; E. Longchar. Attitudes towards botanical growth, interview with A. Deb, January 2, 2022.
6 P. Das. Heritage conservation and plant growth on buildings, interview with A. Deb, December 3, 2021.
7 Dobson and Dempsey (2021) suggest a lack of evidence is one of the root causes of inaction. While here the authors link to inaction with lack of evidence for investment decisions, a more universal lesson is that without the knowledge of values, there is no evidence base for taking action.
8 Bruno Latour argues bringing in democracy into planning in *Politics of Nature* (2004). Along the same lines, Shackleton et al. (2017) point to how planners can engage in democratic planning of green spaces.
9 See the Landscape Advice Note: Vegetation on Walls by English Heritage (Cathersides & Parker, 2014). Habitat potentials with reference to this are discussed in Chapter 5 on Collectives.

10 J. Thoppil, photo-elicitation of landscape fractions with A. Deb, June 7, 2019.
11 J. Thoppil, photo-elicitation of landscape fractions with A. Deb, June 7, 2019.
12 Matthew Gandy discusses this in his film *Natura Urbana: The Brachen of Berlin* (Gandy, 2017).
13 Mugford (2012) in her book chapter discusses the featuring of wild spaces in children's stories.
14 So far, researchers have attempted to understand wild nature itself, but not its connection with urban planning.
15 See Gandy's reference to the sewers of Paris in his book *Fabric of Space: Water Modernity, and the Urban Imagination* (Gandy, 2014, pp. 27–54).
16 E. Longchar. Attitudes towards botanical growth, interview with A. Deb, January 2, 2022.

References

Angel, H. (1980). Guidelines for tree photography. *The International Journal of Urban Forestry*, 4(2), 105–114. doi:https://doi.org/10.1080/03071375.1980.10590599.

Bhalerao, S. (2020, December 20). *Giving trees growing out building walls and highrises new life in the earth*. Retrieved January 3, 2023, from *The Indian Express*: https://indianexpress.com/article/cities/mumbai/giving-trees-growing-out-building-walls-and-highrises-new-life-in-the-earth-7122466/.

Cathersides, A., & Parker, E. (2014). *Landscape Advice Note: Vegetation on Walls*. English Heritage. Retrieved January 3, 2023, from https://www.english-heritage.org.uk/siteassets/home/learn/conservation/gardens-and-landscapes/lan_-_vegetation_on_walls.pdf.

De Sousa, C. (2014). The greening of urban post-industrial landscapes: past practices and emerging trends. *The International Journal of Justice and Sustainability*, 19(10), 1049–1067. doi:https://doi.org/10.1080/13549839.2014.886560.

Del Tredici, P. (2010). Spontaneous urban vegetation: reflections of change in a globalized world. *Nature and Culture*, 5(3), 299–315.

Dobson, J., & Dempsey, N. (2021). Known but not done: how logics of inaction limit the benefits of urban green spaces. *Landscape Research*, 46(3), 390–402. doi:https://doi.org/10.1080/01426397.2020.1864819.

Edensor, T. (2005). *Industrial Ruins: Space, Aesthetics and Materiality* (1st ed.). Oxford: Berg.

Gandy, M. (2014). The Paris sewers and the rationalization of urban space. In *The Fabric of Space: Water Modernity, and the Urban Imagination* (pp. 27–54). Cambridge, MA: MIT Press.

Gandy, M. (2016). Unintentional landscapes. *Landscape Research*, 41(4), 433–440. doi:10.1080/01426397.2016.1156069.

Gandy, M. (Director). (2017). *Natura Urbana: The Brachen of Berlin* [Motion Picture].

Hayes, T. D. (2006). *From slow towns to growth towns: Place representation in a global market*. PhD Thesis, University College London, Faculty of The Built Environment, Bartlett School of Planning, London.
Jorgensen, A., & Tylecote, M. (2007, August). Ambivalent landscapes – wilderness in the urban interstices. *Landscape Research*, 32(4), 443–462.
Leo, C., & Brown, W. (2000). Slow growth and urban development policy. *Journal of Urban Affairs*, 22(2), 193–213. doi:http://dx.doi.org/10.1111/0735-2166.00050.
Loures, L., Burley, J., & Panagopoulos, T. (2011). Postindustrial landscape redevelopment: addressing the past, envisioning the future. *International Journal of Energy and Environment*, 5(5), 714–724.
Mike Raco, D. D. (2018). Slow cities, urban politics and the temporalities of planning: lessons from London. *Environment and Planning C: Politics and Space*, 36(7), 1176–1194. doi:https://doi.org/10.1177/2399654418775105.
Mugford, K. (2012). Nature, nurture; danger, adventure; junkyard, paradise: the role of wildscapes in children's literature. In A. Jorgensen, & R. Keenan (Eds.), *Urban Wildscapes* (pp. 80–96). New York: Routledge.
Otero-Pailos, J. (2016). Experimental preservation. *Places Journal*. Retrieved May 17, 2021, from https://doi.org/10.22269/160913.
Shackleton, C. M., Guild, J., Bromham, B., Impey, S., Jarrett, M., Ngubane, S., & Steijl, K. (2017). How compatible are urban livestock and urban green spaces and trees? An assessment in a medium-sized South African town. *International Journal of Urban Sustainable Development*, 9(3), 243–252. doi:10.1080/19463138.2017.1314968.
Williams, G., Meth, P., & Willis, K. (2014). Grassroots development. In *Geographies of Developing Areas: The Global South in a Changing World* (pp. 321–347). London: Routledge.
Wu, X. (2011). *Productive Landscape: Revitalizing a Post-industrial District with Slow Economy*. Master's thesis, University of Illinois at Urbana-Champaign, Urbana, Illinois.

EPILOGUE

Situating wild spaces in urban nature and its governance

The aftermath of this study should encourage one to look at urban nature as a whole: comprising of both the designed and the wild. It should serve as a reminder that all botanical growths within cities are part of the urban forest, whose behaviour can perhaps be equated to a fragmented ecosystem. Through evidence shown in this book, the fragmented landscapes can be expected to offer a variety of ecosystem services: aesthetics for psychological restoration, extractive benefits of food and fodder, environmental regulation, and biodiversity habitats.[1] The premise has to be set more clearly – I am not suggesting a replacement of designed landscapes – such as parks, swales, allotments for community farming, or rooftop gardens – and turning them into wild spaces. Those designed green spaces are still settings we should strive for. But it would be naïve to not understand that spaces with spontaneous nature also form part of urban nature, and can deliver benefits to help attain urban development goals. Moreover, it is to bring to our own consciousness that microsites of spontaneous nature may be the only settings of nature some neighbourhoods have – and these novel assemblages, for instance plants fused with buildings or wild growth on street sides – are the closest, and sometimes the only, experiences of unintended nature and natural ecological processes to an urban dweller. Wild spaces, much like designed green spaces, if managed well, can deliver ecosystem functions,

DOI: 10.4324/9781003374602-9

that despite their small-scale and fragmented morphology, sum up to become quite significant.

Cities have to confront certain realities. In both Global South and North alike, existing cities cannot be treated the same as brand new canvases even though the key players in land development might hold enough resources to redesign spaces from scratch, because cities already have in place complex multi-layered networks of relationships between people, buildings, infrastructure, and nature.[2] Despite the grim statistics of plummeting green cover in cities, parks and playgrounds do exist and will continue to serve as important typologies of green spaces where nature is experienced by urbanites. However, there are also neighbourhoods without nature. It is a reality that cities are struggling to find resources to support nature. Furthermore, it is to be equally regarded that allowing spontaneous nature to grow (or to reclaim the urban spaces) without some kind of control can be problematic: i.e., wild spaces also carry negative externalities (such as affecting human health and damaging infrastructure).[3] There also emerges a concern that if spontaneity is embraced, then the city might become complacent in planning and designing of green spaces. This is problematic because without planning or designing for the future, meeting the changing needs of society will be difficult to achieve. However, despite reasons to do away with spontaneous nature in cities, they continue to exist due to a confluence of factors outside of human control.

It is well acknowledged among urbanists how planned spaces for nature in cities are not always realised in the same manner as they are envisioned.[4] For instance, spaces allotted along the roadsides for green infrastructure purposes (such as grass laden bioswales, shrubs for aesthetic visuals, or tree lined avenues for shade), may be devoid of nature in reality because of a shortage in resources and labour of the public department responsible for implementing these visions. As another example, private gardens envisaged in development control are not kept by landowners but are instead built upon due to demand for space – for personal lifestyle preferences. Or consider a scenario where real estate actors who depend on rent and housing sales decide to trade provisioning of green spaces for more housing units to meet market demand in response to urban population expansion. Given the myriad of ways planned spaces are, or are not, manifested, the delivery of green spaces can then be understood as a complex affair. Yet, nature grows in cities beyond the ways that had been imagined: sometimes nature grows in neglected spaces such as street sides and parks, and other times, nature grows even over buildings and built-surfaces which were not supposed to be green spaces. Several peculiar examples discussed in this book include residual spaces around street utilities, rooftops of buildings, and inside storm water drains.

Urban spaces that are reclaimed by plants to counter the apparent disbalance humans have created in cities by removing nature, compels one to revisit James Lovelock's hypothesis of *Gaia* (Lovelock, 2000). Given the current state of urban affairs, wild growth created by forces of Gaia offer a distinct window of opportunity. Despite the importance of land use plans for an orderly future of cities, their perceptions may take up to 15 years or more to show effect on ground. Moreover, plans may not materialise at all. That means communities can go on for decades without any signs of neighbourhood renewal or revitalisation. In the interim, if having "some" green is better than having "no" green,[5] then it is certainly worthwhile to explore how wild spaces can be retained and exploited for urban development.

Although wild growth happens outside human control, they continue to actively or passively play a role in the everyday life of societies around them. Particularly, wild growth which is small in scale are more ubiquitous, and are hard to exclude from our everyday experiences: it would be rather hard, if not impossible, to find a neighbourhood without grass growing from the cracks of its pavements. Unintended settings such as trees growing upon buildings however, are typically difficult to find in new neighbourhoods; yet they are a common sight in neighbourhoods that comprise an ageing, built fabric of the city. Relatively easier to spot might perhaps be parks that have wild growth instead of designed nature; such is commonplace in neighbourhoods facing austerity cuts. Plants are also very quick to colonise buildings that have been left half completed or are abandoned. Despite their pervasive presence, the lives of and lives around these natures are seldom noticed, due to the smallness in scale. But they have without our conscious, become part of everyday life, adding to visual aesthetics and supporting a wide variety of animal life.

Perhaps it is useful to note here that in many ways, wild spaces make cities inclusive for both human beings and non-human life. The wild spaces are less discussed in terms of the benefits or harm it causes to animal life in cities. The animal component in cities is often overlooked, but they have a link with human beings as well as wild spaces. Some of these useful links are domesticated cattle that graze on wild spaces and in turn offer food products, or more autonomous lifeforms like birds which are a treat to the eyes, or wild flowers attracting bees and other pollinators which are important for plant growth in cities. Yet, people who walk by these assemblages of wild nature (or microsites of botanical growth) tend to ignore them while attending to their busy schedule of the day, and in turn end up disregarding these interesting and fascinating spaces as everyday nature. However, the streets would be empty and devoid of life when such assemblages go missing – cattle, flowers, butterflies, ladybugs, birds, and other non-human life that made for a vibrant ecology, may no longer be found.

One would hope that perhaps their values would be felt more in their absence. But often the removal of these landscapes also goes unnoticed. In one such incident, a street upgrading project in my neighbourhood was set to remove not only the existing ground cover, but would pour concrete over the entire right of way, sealing the surface, and leaving no space for new plants to grow. Notwithstanding, the public authorities had made some considerations for existing trees. Upon discussion with site engineers who were supervising implementation of the street designs I felt that they well understood the value of plants for the wellbeing of the neighbourhood. However, the top-down decision making spared little opportunity for the feedback which can stop the construction in time. In less than a week the roadside verges that earlier served space for wild growth had now disappeared. Neighbourhood dwellers, however, did not find this the most pressing issue as they found the aesthetic of concrete roads to be cleaner – without realising we have removed food for cattle, have obliterated an important layer of habitat for pollinators, and done away with a supplement to the drainage system for checking rainwater runoff that now does not exist. This points to the fact that it is not just a problem of how projects are bureaucratically designed and implemented, but also to the fact that citizens (including myself at times) are disposed to an aesthetic that is devoid of nature. Having said that, city makers would also have to appreciate that some citizens actually enjoy the messy aesthetics of spontaneous nature that draw animals, birds, and insects to create a wholesome experience of nature. There is beauty associated with naturalistic landscapes.

Cities must have beautiful green landscapes, for indeed, that is one of the requisites of good quality urban life; consequently, also the focus of urban designers and planners. Visible change – particularly when neighbourhoods go through an aesthetic overhaul for facelift – can improve residents' image of their own community. As a resident, I too felt good when the city authorities rebuilt the public roads within my neighbourhood – it made the place look new and different. I had cherished the old neighbourhood landscape as well, where there was green wild growth on either side of the streets. Perhaps it was the novelty of change, however momentary, that brought a fresh outlook to the place which made me feel good. At this point, it is worth revisiting how beauty is not the only value of urban greens, and to understand the value of wild spaces, one needs to account for other critical services which these ecosystems offer. I would have argued the same within my community mentioning how wild plants help reduce pressure on public drainage infrastructure in slowing storm water runoff. A few months ago, even the little service which the narrow strips of green can offer would have sounded perfectly important as the street was notorious for waterlogging during heavy rains. Residents were growing tired of rainwater flooding their homes and were willing to consider any measure that could provide even the slightest of relief. However, more recently, the public drainage was upgraded and waterlogging is no longer a problem – for now. It is unlikely that I will be able to convince my community of the benefits of green infrastructure for a faraway future. Yet it is in the future when the now low-rise low-density quaint neighbourhood in the city's outskirts will transform into a bustling district – exerting more pressure on the public drainage system.

Given how little wild spaces matter in planning and design at present, there needs to be a more authoritative approach. Evidence can play a big role in elevating the dialogue and offering wild growth the much-needed space within discourses of urban greening. To appreciate the value of wild growth, there needs to be a thorough understanding of these assemblages. For those of us who may not have an understanding of the ecosystem services that urban green spaces offer beyond visual beautification, it is certainly hard to critique the value of wild growth in its entirety. Moreover, in this book I am discussing microsites in cities – a class of natural growth that has been far less represented even in theory. It is not surprising, then, when people do not value small-scale wild spaces in cities. There has also been an emerging stream of discourse that looks at values of wild landscapes. Such discourses encourage cities to preserve wild growth instead of dismissing them as weeds or undesirable nature. However, to look at spontaneous nature as a mode of urban greening, it is perhaps necessary to further understand which assemblages can be integrated meaningfully into everyday settings and how cities can go about doing so. Not all nature is desirable, but at the same time, utility of nature can be improved through design measures. To illustrate, this book has looked at a few methods through which urban dwellers have integrated small-scale assemblages of spontaneous nature into their lives (see Chapter 6 on Co-production).

The focus on wild nature has seen a sort of increase since 2000. Even during that short stint of research that has focused on wild spaces in cities, the sub-field has managed to reflect a certain saturation in the knowledge and discourses surrounding wild spaces. While there is a lot of discussion on biophysical aspects, the social lives of these spaces are less explored. Particularly, the existence of wild nature that is small-scale has almost entirely remained absent from the dialogue. This is scary – because it shows how upscaling has become the norm and how we are unwilling to think about smaller arrangements that constitute everyday life. In the attempt for cities to achieve global goals of sustainability, the small-scale assemblages so far, seem to command little or no attention. Even biologists, who are among the first guardians of nature, at times seem convinced that there is little legitimacy in exploring wild urban settings – they consider that strength of the negative discourses far outweigh the positive ones.

Particularly, the focus of this book on small-scale wild growth addresses an additional problem of such nature being invisible and being further ignored as they may serve very little utilitarian purpose. Moreover, unlike the large-scale wild spaces – for instance a riparian vegetation along a stream passing through the city – small-scale nature is more closely entangled with the built-form of the city leading to several risks to inhabitants and users. Yet a more careful review of these assemblages reveals that certain fractions of spontaneous nature might be worth preserving. Benefits of landscape fractions are rather perceptible to the eye, i.e., when they are aesthetic; when they support biodiversity habitats; when they support urban grazing; and so on. Chapter 5 on Collectives should encourage readers to look at small-scale assemblages of spontaneous spaces as networks rather than as isolated fractions. These networks may be imagined at the street level, neighbourhood level, or even at the city level. Other potential benefits towards sustainable development such as the links to human health and climate regulations are clearly there, but more empirical research is needed to understand the magnitude of the contribution.

Considering the circumstances, among the things that would be interesting to see, for me at least, is how cities would use the ideas in the book to experiment with policies and practices that help manage wild urban spaces in a manner that can help achieve a healthier and greener neighbourhood. The potential for using spontaneous nature as a mode of greening appears to exist both for developing and developed countries alike. In fact, presently many cities, particularly European countries have adopted such policies and management practices that help integrate wild nature into everyday green spaces. Yet there are also cities elsewhere which have strict controls to prevent spontaneous nature from cropping up. Then there are cities that have no mention of wild spaces in the urban greening policies at all, as if they do not exist. The potential of assemblages I discuss as landscape fractions, would depend on how cities decide to respond to such natures. If cities are indifferent, they will continue to exist without any enhancement to their intrinsic potential. If these spaces are co-designed with the purpose of integrating them more meaningfully into the society, then these assemblages would emit some more value, in economy, environment, and socio-cultural aspects. If cities decide to treat them as unwanted objects, they will likely disappear despite their strong instincts and circumstances to survive.

Whether or not the value of small-scale wild spaces is enough to sway people will depend on how creatively these landscape fractions are used and managed. There should, however, be no doubts about the impact small improvements can make to the quality of life in the neighbourhood; to prove this, ample examples exist labelled "urban acupuncture", "do-it-yourself (DIY) installations", "tactical urbanism", and so on.[6] Having said that, there are obvious advantages that come with large green spaces that are meticulously designed by professionals to cater to certain – and often innovative – utilitarian purposes: bioswales, avenues, boulevards are some tools of the landscape architects to uplift the sustainability and resilience of the built environment. However, looking at the extent to which cities have failed to provide nature as part of people's everyday experiences, managing designed landscaping and managing wild growth can be seen as simultaneous goals rather than competing goals.

Notes

1 See Chapter 5 on Collectives for discussion on ecosystem services.
2 See Swyngedouw (2006) on networked technostructures, discussed in Chapter 2 on Meanings.
3 See Chapter 2 on Meanings and Chapter 3 on Frames for negative discourses.
4 See Lefebvre (1991) for discussion on perceived and conceived space.
5 See Wolch, Byrne, and Newell (2014) on their proposed approach of "just green enough".
6 See Arefi and Kickert (2019) and Loder (2020) for discussions and proof of these concepts.

References

Arefi, M., & Kickert, C. (2019). *The Palgrave Handbook of Bottom-up Urbanism*. Cham, Switzerland: Palgrave Macmillan.

Lefebvre, H. (1991). *The Production of Space*. (3rd ed. Translated by Donald Nicholson-Smith. Cambridge, MA: Basil Blackwell.

Loder, A. (2020). *Small-scale Urban Greening: Creating Places of Health, Creativity, and Ecological Sustainability*. New York: Routledge.

Lovelock, J. (2000). *Gaia: A New Look at Life on Earth*. New York: Oxford University Press.

Swyngedouw, E. (2006). Circulations and metabolisms: (hybrid) natures and (cyborg) cities. *Science as Culture* 15(2): 105–121.

Wolch, J. R., Byrne, J., & Newell, J. P. (2014). Urban green space, public health, and environmental justice: the challenge of making cities 'just green enough'. *Landscape and Urban Planning* 125: 234–244.

INDEX

Abandoned, xvi, xviii, xix, xxvi, 4, 22–23, 26, 29, 33, 36–37, *50*, 54, 66–67, 75–76, 82, 86n12, 96, 99, 110–111, *113*, 118, 143, 148, 151, 162–163, 167, 176 (see also: decommissioned, dilapidation)
Abandonment, 14, 29, 67, 110–111
Adaptation, xxii, 112
Addis Ababa, xiii
Aesthetic valuation, 69
Aesthetics, xvi, xxii, 3, 12, 16, 18, 22–24, 34, 38, 43, 58, 62, 65–88, 90, 112, 115, 118, 120, 129, 135, 139, 143, 147, 153n4, 157, 161, 165–166, 169, 170, 174, 176, 177
Agencies, 2, 43, 54, 61, 119–120
Agency, 47, 52, 65–66, 113–114
Ahmedabad, 94
Algae, 52, 74, 100, 122–*123*, 144
Allergic, 37
Allergies, 66, 145, 161, 171
Alley, 9, 11, *32*–33, 138 (see also: narrow streets)
Ambivalence, xxiii, 14, 27, 159, 161–162
Ambivalent landscapes, 121
Amenities, 95, 158
Amherst Street, 17n11, 121, 146–148
Animals, 6, 8, 25, 36–37, 62n5, 98–100, 144, 149, 159, 166–167, 177
Anthropocentric, 161
Anthropogenic, 43, 47, 58, 66, 169
Arboriculturist, 132
Architects, xiv, xxv–xxvii, 3–4, 12, 25, 70, 71, 73, 85, 98, 104, 135, 138, 142, 164, 168–169, 180
Architectural elements, 23, 67, 75, 136, 141
Architectural legacies, 23, 82, 160
Architecture heavy, xxii, 73–74
Art nouveau, 70, 85n6
Art value, 69, 71, *147*, 165
Artificial, 28, 66, 71–72, 124
Association of plants, ix, *57–58*
Autonomous, 161, 171n3, 177

Backyard, xiv–xv, xvii, 1, 29, 65, 71, 94–95, 156, 167
Banyan, 10, 31, 51, *55*, *80*–81, 146
Beatley, Timothy, xxiv, 15, 16
Beautification, xviii, xxviii, 67, 83, 142, 178
Bengaluru, xv, xx–xxi, xxvi, xxviii, xxix–xxx, xxxi-n4, 3, 5, *7–9*, 17n7,

17n8, 21, 23–*24*, 33, 46, 49, 51, *52–54*, *56–60*, 63n9, 67, 70, 81–*82*, 91–*93*, 95, *99*, 107n4, 107n5, 114, 118, 121–*122*, *123*, 130, 146, *149*, *150*, 163 (see also: KR Market, Sandal Soap Factory)
Berries, 1, 94, 100, *104*, 161
Biofilm, 63, *123*
Biological growth, xix, xxii, xxix, 3, 11, 17n11, 26, 28, 31, 43–*44*, 49, 52, 55, 58–59, 75, 100, 145–146, 163, 167, 170
Biophilia, 38n6, 98
Biophilic patterns, 72
Bioswales, 175, 180
Bird droppings, 31
Birds, 29, 35, 38n6, 65, 98–100, 103, 124, 145, 149, 159, 167, 177
Blight, xxi, xxiii, 29, 94, 147
Botanical growth, xvi, xxix–xxxi, 17n12, 22, 25–26, 47, 50, 57, 67, 103, 106, 107n7, 113–114, 127, 136, 140, 148, 171n5, 172n16, 174, 177
Botanist, 9, 15, 46, 120
Bottom-up approach, xv, 166
Braun-Blanquet, J., 46, 57
Buenos Aires, xiii, xxxi
Built environment, 3, 13, 27, 30, 32, 42, 107, 151–152, 158
Bursts of green, 115, 129
Business, xxvi, xxviii, 15, 48, 51, 81, 91, 157
Butterflies, 5, 7, 9, 100, 103, 115, 146, 177

Cattle, xix, 35, 37, 91–92, 94, 118, 158, 177
Chihuahua, xxxi
Chikunguniya, 103
China, 151, 165
Circulation, 25–26, 38n4, 47, 64
Citizen-led, 153
City centre, xxix–xxx, 124
City master plan, 148
Civic movement, 130
Climate, 4, 13, 16, 54, 57, 60, 67, 75, 95, 98, 103, 148, 159, 171n1, 180–182, 188, 233, 256, 257, 264, 271, 273, 291 (see also: Climate change)
Climate change, xxxin2, 16, 26, 46, 104–109
Climbers, *104*, 114, 144
Coal yard, *82*–83
Co-creation, 112
Co-existence, 163, 165, 169
Collective memories, xix, 78
Collectives, xxii, 35, 37, 38n6, 69, 83, 89–109, 120, 154, 157, 165–166, 170, 171n9, 179, 181n1
Colonial, xxvi, 10, 23, 49, *50*–51, 54, 64, 70–71, 75, 79, 83, 85, 89, 108, 112, 127, 139, 148, 160
Community-led, xx, 129
Connectors, 9, 166
Conservation architect, xxvi, 144, 146, 164, 168
Conservationist, xiv, 59, 135, 162
Control box, xvi, 7–9, *59–60*, 130–*131*, *149*, 150, 165, 167
Convenience, xxi, 45–46, 62, 77
Cool, 32, 105, 132. 167
Cooling, 35, 51, 83, 105, 108
Co-production, xix–xx, xxii, 63, 67, 110–133, 165, 179
Counter-utilitarian, *44*, 45, 66, 96
Crabs, 74
Crabbing, 74
Cuba, 34, 39
Cultural, xx, xxii, 23, 35, 58, 69, 78–79, 83, 103, 106, 112, 161, 163–165, 169, 180
Cyborg, xx–xxi, xxviii, 24–26, 28, 38n3, 64

Damage, 15–16, 36, 55, 66, 94, 124, 134, 143–144
Death, 27–28, *30*, 168 (see also: urban decay)
Deaths, 106, 103
Decommissioned, 4, 36, 83, 112, 151
Dee, Catherine, 86n13, 138,
Demand for space, 158, 175 (see also: scarcity of space, low urban green space)
Democracy, 113, 133, 171n8
Democratic planning, 171n8
Dempsey, Nicola, xxv, 16, 171n7, 172

Dengue, 48, 103
Deprived neighbourhoods, 167 (see also: low-income neighbourhoods)
Derelict, xviii, 12, 14, 27, 29, 69, 77, 106, 110, 121, 145, 159, 164–165
Design, xv, xxii, 1, 16, 27, 73–74, 83, 85n6, 118–120, 127, 129, 136, 142, 167, 170, 175, 178–179 (*inter alia* re-design, urban design)
Detritus, 54
Developers, 95, 148, 158, 169
Development control, 175
De-weeding, 2, 98
Dilapidation, xviii, 60, 139, 144–145, 171
Dirty, 146–147 (see also: untidy, messy)
Disease, 69, 103, 143, 145, 161
Do-it-yourself (DIY) installations, 180
Dover Castle in the UK, 146
Drainage, xxix, 36, *44*, 55, 70, 105, 177–178
Drosscapes in Germany, 86
Dry, 21, 31, 54, 100, 115, 119, 141, 149, 153n1, 157

Eco-centric, 161
Ecological succession, xxiii, 29, 60, 111–112, 127, 132n1
Economic change, 112
Economy, xxviii, xxix–xxx, 33, 35, 43, 50, 70, 81, 90, 95, 145, 158, 166, 180 See livelihood
Ecosystem services, xix–xx, xxii, 15, 91, 106, 145, 159, 168, 174, 178, 181n1
Education, xx, 6, 15, 29, 122, 148, 161, 170
Effluent treatment tank, 122
El Ninõ, 48
Emission, 47, 82, 104
Engineers, xx, 9, 42, 44, 132n7, 135, 138, 144, 163, 170, 177
English Heritage, 100, 124, 144, 154n12, 154n17, 171n9
Entrepreneurship, 162
Environmental goals, 27 (see also: SDG, sustainable development goals)
Ephemeral growth, 55, 148, 160
Evapotranspiration, 46, 105, 153n1, 167
Everyday experiences, 98, 127, 167, 176, 180
Everyday nature, xvii, 15, 33, *80*–81, 89, 151, 154n19, 158, 167, 177
Experiences of nature, 65
Expert-driven, 130
Exposed masonry, xviii, 72, 129, 137

Factory, xxvi, *9–11*, 17n2, 22, 54, 57, 63n7, 81–*82*, *84*, 100, 122, 123, 163
Fibonacci progression, 72
Fig tree, xxix, 13, 41, 135
Films, xx, 71, 122, 163, 169
Film-soils, 52–54
Fiscal transfers, 160
Flash landscape, 55–57
Fletcher, Banister F., 85
Flowers, xx, xxviii, 1, 5, 7, 9, 28, 35–37, 45, 54, 57, 72, 85n6, 91, 94, 100, 103, 115, 124, *131*, 132, 141, 145–146, 151, 165, 177
Fodder, 37, 91, 106, 174
Food, xviii, xx, *34*–35, 90–92, *99*–100, 103, 128–**129**, 166, 176–177
Foraging, xx, 65, 91, 94–95, 157, 161
Formalism, 71–72
Forman, Richard T.T., 5
Foucault, Michel, xxiv, 14, 27, 38n5, 46, 59, 62n5, 154n10
Foul smell, 69
Fractal Geometry, 72
Fragmented, xix, 72, 90–91, 106, 152, 160, 166, 174–175
Frame Analysis, 153n5
Framing, 43, 135–136, 138, 141, 148, 150
Frogs, 9, 65, 98

Gaia, *30*, 110–111, 113. 115, 125, 159, 171, 176
Gandy, Matthew, xxiv, 1, 10, 14, 16n1, 22, 42–43, 45, 63, 66, 69, 71, 80, 86, 96, 103, 108, 135, 152, 162, 172n12, 172n13, 172n15
Ganges, 77, 89, 145

Garbage dump, 146, 147,
Gardens, xv, xx, 3, 10, 28, 31, 33, 62n3, 69, 95–96, 104, 174–175
Ghost walks, 37
Gillian Rose, 73
Glenn Parsons, 65
Global capital, 47
Global south, xiii, 13, 19, 158, 175
Globalisation, xxx, 12–13, 43, 157, 159
Golden proportion, 73, 136
Governance, xiv, xxiii, xxviii, 4, 174
Graffiti, 21, 48–*49*, 67, 111, 165
Grasses, xiv, xxix, 2, 7, 9–10, 14, 35, 41, 46, 54, 62, 94–95, 98–100, 103, 115, 124, 153n1 (see also: ground cover)
Grassroots, xx, 157–158, 171n3
Grazing, xix, xxii, 35, 91–94, 106, *117*–118, 121, 132n5, 132n6 (see also: cattle)
Green infrastructure, xv, xxi, xxxii, 26, 31, 62n3, 95, 156–157, 167
Green spaces, xiv, xv, xxi, xxiii, xxvii, xxx, xxxi-n1, 2, 10, 15, 17n2, 17n3, 17n4, 33, 35, 154n16, 159, 170, 175
Greening, xiii, xix, xxi, xxix, 3, 26, 31–33, 35, 62n1, 80, 95, 98, 115, 127, 133, 142, 157, 158, 159, 165, 166, 168–169, 171–172, 178–179, 180
Greening project, xiii
Ground cover, 17n6, 31, 57, 98, 105, *117*, 124, 177
Guerrilla gardening, 34

Habitat, xviii, xxvii, 5, 9–10, 13, 31, 36–37, 46, 69, 83, 98–103, 106, 118, 122–*123*, 137, 143–144, 159, 161, 163, 166–167, 171n1, 171n2, 171n9, 174, 177, 179
Haraway, Donna, xxiv, 24–25
Harmful, xxxi-n2, 37–38, 45, 125, 161
Harmless, 38, 124, 147
Haunted walks, 163 (see also: Ghost walks)
Heating, 102
Henri Lefebvre, xxiv, 13, 17n15, 150, 154n20, 181n4
Herb, 37, 94, 114
Herbaceous plants, 99, 105, 144, 153n1
Heritage, xviii, xix, xx, xxii, xxvi, xxvii, 4, 21, 23, 49, *55*, *58*, *75*, 78–79, 82–83, 100, 123–124, 125, 142, 144, 146, 153, 154n11, 154n12, 154n14, 154n17, 154n18, 159–160, 164–165, 169–170, 171n5, 171n6, 171n9 (see also: English Heritage)
Herons, *99* (see also: birds)
Hidden, xiv, xvii, xxiii, 7–9, 26, 31, 37, 148, 150, 152, 159, 160, 170
Hide and seek, 37
High Line project in New York, 83, 86n14
Historic buildings, 22, 144
History of Architecture, 85n6
Homemakers, 92, 94
Hong Kong, xxxin1, 28, 55, 144, 163
Human-dominated systems, xxi, 43, 45, 62n3
Hybrid landscapes, 129, 130, 132n7
Hyderabad, xii, xxxi-n1

Iconic buildings, 78
Identity, xxi, *57*, *58*, 159
Illogic of space, 163
Imagination, xviii, xxiv, 2, 14, 37, 67, 71–72, 75–76, 91, 110, 152, 162, 169, 172n15
Immunity, 103
Inaction, 171n7
Industrial legacy, 23, 57, 81
Industrial site, xxii, *82*, *84*, 112, 162
Injurious, 36
Innovation, 162
Insect kingdoms, 165
Insects, 9, 13, 36, 94, 98–100, 103, 111, *123*, 145, 151, 159–161, 167
Instagram, 69
Interconnected, 18n16, 90
Invasive, 5–6, 10, 37, 100, 132, 165, 170
Investment, xviii, 91, 171n7
Island Laboratories, 166

Jakarta, 47
Jorgensen, Anna, xxiv, 1, 14, 38n7, 85n4, 86n13, 96, 121, 159, 162,
Just Green Enough, 33, 98, 104

Kaplan, Rachel, 38n6, 69
Kolkata, xv, xxi, xxiv, xxvi, 3, 10–*11*, 17, 21–23, *30*–34, 49, *50*–51, 54, 62–64, 70, 74–79, 81, 89, 92, 99, 110–*113*, 121, 127–*128*, *130*, 137–148 (see also: Amherst Street, Kumartuli)
KR market, xxviii–xxx, *53*
Krzywoszynska, Anna, 52
Kumartuli, 49, 89–*90*, 145,

Ladders of growth, 157
Landmark, 142, 169
Landowner, xxix, 2, 31, 45, 62, 121, 147, 150, 162, 170, 175
Landscape-ecologist, 5
Landscape fraction, xv–xvi, xviii–xxiii, xxv, 3, *5*–7, 14–15, 22–23, 25–28, 30, 32–33, 35, 38, 45, 47–48, 55, 59, *61*, 63, 66, 70, 72–73, 77–78, 80, 82–83, 85, 86n10, 86n11, 96, 103, 105–106, 113–114, 119–120, 132, 135–136, 141–142, 148, 150–153, 159–162, 166–169, 172n10, 172n11, 179, 180
Landscape managers, 124, 145–146
Landscape transformations, 119
Landschaftspark in Germany, 82
Latour, Bruno, xxiv, 16, 47, 62n4, 113, 171n8
Leaking pipes, xxix, 46, 51
Legal, 10, 45, 54 (see also: illegal)
Life-cycle, 52, 135
Leopold, Aldo, 85
Litter, xxii, xxviii, 47, 52, 120, 130–*131*, 147, 149, 157, 167
Liveable, 4, 237, 253
Livelihood, 90–91, 106, 121, 157
Livestock, xviii, 91–93, 109, 118, 166
Living map, 148
Living organism, 111, 123
Local government, 42, 89–90, 115, 172
Lorimer, Jamie, 103, 108
Lovelock, James, 30, 39, 111, 133, 176, 181
Low urban green space, xxxi
Low-income neighbourhood, xxi, 4, 142

Malaria, 48, 103, 143
Manage overgrowth, 115
Management regimes, xix, 157
Managing biological growth, xxii, 17n11, 40, 163, 167
Manchester, 105
Mandelbrot, Benoit B., 72
Marginal spaces, xx, 14, 33–34, 66
Masdar Institute in Abu Dhabi, 81
Meanings, xviii–xxi, xxiii, xxvii, xxxi, 14–15, 21–40, 69, 77, 96, 104, 118, 120–121, 136, 138, 151, 153n7, 159, 165, 167–168, 181n2, 181n3
Memory values, 83, 135, 170
Mercantile, 89–90, 145
Messy, xxviii, 98, 169, 177
Metabolism, 25–26, 38n4, 47
Metropolitan, xiii, 17n5, 93
Microbes, 103
Micro-ecologies, 103
Microhabitats, xvi, 122 (see also: habitats)
Microorganisms, xxxin5, 103
Microsites, *5*–6, 23, 33, 37, 48, 96, 125, 211, 246, 248, 273, 284, 287, 289
Migration, xviii–xxi, 14, 18n16, 43, 49–50, 78, 89
Mill, John Stuart, 64, 87
Mode of greening, xxi, *32*, 98, 115, 127, 157, 159, 168–169
Moisture, 36, 45–46, 52, 143
Monument, xxii, 23–24, 77–78, 110, 160, 164, 169–170
Moods and feelings, 86n9
Mosquitoes, 48, 103, 108, 109
Moss, xvi, 21–22, 28, *32*–33, 35, 45, 52, *57*, 62n5, 78, 100, 114, 146
Mumbai, xiii, xxxi-n1, 163
Mushrooms, 94

Nadar, Félix, 152
Narrow streets, xxix, 32, 127, 129 (see also: alley)
Natural ecosystems, 5, 30, 112
Naturalist, xxiv, 2, 38n6, 85n8, 134
Naturalistic, 42, 145, 177
Naturalness, xix, 14, 71, 96, 98, 146
Nature walks, 153, 169
Neglect, xiii–xix, xxix, xxx, 4–7, 30–31, 33, 43–45, 48, 54, 61, 70, 76,

82, 92, –93, 103, 111, 114–115, 118, 124, 141, 157, 159, 167, 169–170, 176
Negotiating power, 47
Netherlands, xxvii, 121, 132n5, 132n6
Networked technostructures, 25–26, 181n2
New Urban Agenda of United Nations, 156
Nimelä, Jari, 120
Non-design, 14, 62, 170
Non-human life, 25, 35, 91, 106, 163, 166, 176, 177
Non-profits, 165
North American countries, 165
Novel ecosystems, 42, 67, 132n2

Obsolete, xviii, xxix, 78, 81, 112
Old city, xviii, xx, xxvi, xxix, 4, 34, 50,
Organopónicos system, 34
Outside human control, xiv, 43, 107, 171, 175
Oversight, xvii, 111, 139, 147, 159
Oxford, 145

Paint, xx, xxix, 23, 132 (see also: worn-out paint)
Paris, 172n15
Park, xv, 5–6, 17n7, 92, *104*, 114–115, 119, 132n7, 158
Parthenium, 6–7, 10, 37–38, 78
Pastoral, 93
Per capita green open space, xxxi
Physiological, xxxi-n2, 54, 96, 174
Placelessness, 110
Places of decay, 169
Place-making, xix, xxvii–xxviii, 128, 160, 166–167
Plant borne-allergens, 69
Plant sociology, 45–46
Planter box, xiii, xxx, *49, 128*, 138
Playgrounds, xxi, xxx, 2, 37, 157, 175
Plaza Republica, 48, *49*,
Political ecology, xxiv, 113
Pollution, xxxi-n2, 83, 104, 106
Poor neighbourhood, xiv
Post-industrial, 23, 67, 81–*84*, 160, *164*–165
Precarious activities, 37
Production of space, xxiv, 13, 17n15, 43, 129, 150
Pro-environmental attitudes, 16, 27
Property owners, xix, 41, 144
Pruning, 105, 114, 121
Public authorities, xviii, 48, 92, 115, 118, 147, 177
Public health, xxvii, 103, 135, 148
Public infrastructure, 31
Pumpkin, 35, 94

Quito, xv, xxvii, 3, 48–*49, 76–77*

Rainwater, 32, 36, *44*, 52, 55, 70, 143, 177, 178
Rajarhat, 50
Ramdulal Sircar Street, *30*, 138,
Reclaimed landscape, 112, 114, 119
Red brick masonry, *89–90*, 111, 138
Regrowth, 59,
Reigl, Alois, 22, 73, 86n9
Relationship between nature and human, 165
Relph, Edward, 110
Repair, 4, 42, 51, 98, 124
Repoussoir, 138–*139*
Repurposing and redesigning, 85
Risk, xiv, xviii–xx, 2, 36–37, 51, 54, 83, 94, 98, 103–105, 114–115, 121, 130, 141–*143*, 145–146, 152, 159, 161–165, 168, 171, 179
River Esk, 74, 124
Roof, 3, 10, 13, 36, 104, 111, 138, *143*, 152, 156
Root action, 35, 69
Roots, xx, xxviii, 10, 13, 25, 43, 51, *57*, 110, 113, 121, 135, 140, 144
Ruins, 27, *76*, 110

Salt Lake City, 50
Sandal Soap factory, xxvi, 63n7, 81
Saplings, 28, 42, 94, 104
Scarcity of space, 128 (see also: demand for space, low urban green space)
SDG, 156–157 (see also: sustainable development goals)

Seasons, 21, 118, **129**, 149, 169–170
Sensitization of the local communities, 153
Sensory enframement, 66,135
Shanghai, xiii, xxxi-n1
Sheffield city, 21, 42
Shrub, xiv, 1, 2, 7, 10, 14, 17n6, 29, 31, 52, 59–62, 72, 78, 95, 105, 115, 121, 127, 141, 175
Slow growth areas, 158
Small-urban places, **128**
Snake, 6, 9, 36, 37, 65, 161
Social change, 93
Social lives, 151–152, 179
Social networking, 69
Socio-ecological, 47
Soil, xvii, xxiv–xxx, 13, 23, 26, 31, 35–36, 41–42, 46, 52, 54–56, 81, 92, 94, 99, 106, 165 (see also: film soils, Soil Care Network)
Soil Care Network, 52, 63n8
Status, xiv, xxiii, 66, 72, 136, 141, 145–146, 168, 170
Stepping stones, xxvii, 98–99, 103, 106, 166–167
Stigma, xviii, 38
Stinging nettle, 37
Stormwater drain, *150*
Straggler daisy, 132
Stranglers, xxix, 10–11, 35, 51–52, 110, *113*, 138–139, 144
Streetscape, 95–96, 98, *117*, 141
Streetside, *60, 93*, 95
Supervenience, 85
Surveillance, 103
Sustainable development goals, 157, 159
Swiss Re Tower in London, 81
Swyngedouw, Erik, xxiv, 25, 38n4, 47–48, 103, 181n2

Ta Prohm, a 12th century temple in Cambodia, 144
Tactical urbanism, 180
Tallinn in Estonia, 120
Technological relic, xvi, 7–8, 29, 167
Temporal, 29, 59, 91, 113, 115, 141, 148–149, 169, 173
Texture, 29, *137*–138, 141
The Good Eye, 73, 87,
The University of Sheffield, xxiv–xxvii, 52
Top-down, xv, xxi, 13, 118, 129–130, 149, 177
Topiary, 128
Tourism, xx
Town planning regimes, 92
Trajectories of wild urban spaces, 112
Trajectory of abandoned spaces, 86
Transience, 60
Transitory, 11, 150, 159, 169
Tredici, Peter Del, xxiv, 15, 27, 33, 96, 120, 165
Tree care, 51
Tree failures, 55
Tree photography, 134, 153, 154, 172
Trellis, 137, 146

UHI, 105–106 (see also: urban heat island)
Unhealthy, 45
Unintended, 1, 5, 43,66, *68*, 80, 174, 176
Unintentional landscapes, 43, 66, 69, 80, 108, 165
Unmanaged growth, 130
Unplanned, xvii, xxv, 8, 25, 31, 42, 45, 158, 170
Untidy, 169 (see also: messy, dirty)
Urban acupuncture, 180
Urban decay, xviii, xxiii, 27, 29, 48, 66, 121, 148 (see also: places of decay)
Urban environment, xiii, xx, 47, 69, 104, 138, 156, 171n2
Urban expansion, 31
Urban forest, xiii–xv, xxi–xxii, 2, 13, 17n5, 41, 62, 64, 105–106, 109, 174
Urban heat island, 105 (see also: UHI)
Urban periphery, 17n8
Urban planning, xx, xxxii, 17, 18, 20, 35, 39, 40, 107n2, 107n3, 107n7, 118, 172n14
Urbanisation, xiii, xxi, 15, 18n16, 30, 79, 94, 170

Vandalism, xxiii, 70, 130
Vertical surfaces, 10–*11*
Vistas, xxiii, 129, 136
Visual appreciation, 66, 136, 145
Visual directness, 86
Visual Methodologies, 73
Visual relief, 31–*32*, 35, 52, 167
Visual tranquil, 98

Waste management, 92
Wasteland, 14, 48, 69, 71, 159, 165
Water supply, *8*–9, 26, 31, 60
Waterlogging, 36, 123, 178
Weeds, xviii, 14, 100, 146, 152, 179
Whitby, xv, xxvii, 3, 63n7, 73–*74*, 123, *125*, 127, 146
White topping, 98, 118
WHO, xiv, 13
Wild Urban Woodlands, 132n4
Worn-out paint, xxix, 130, 139
Worn-out plaster, 90

Zangwill, Nick, 85

For Product Safety Concerns and Information please contact our EU representative GPSR@taylorandfrancis.com
Taylor & Francis Verlag GmbH, Kaufingerstraße 24, 80331 München, Germany

www.ingramcontent.com/pod-product-compliance
Ingram Content Group UK Ltd.
Pitfield, Milton Keynes, MK11 3LW, UK
UKHW022344190926
475702UK00015B/2097
9781032449333